Counterfactuals and Probability

Counterfactuals and Probability

Moritz Schulz

OXFORD
UNIVERSITY PRESS

OXFORD
UNIVERSITY PRESS

Great Clarendon Street, Oxford, OX2 6DP,
United Kingdom

Oxford University Press is a department of the University of Oxford.
It furthers the University's objective of excellence in research, scholarship, and education by publishing worldwide. Oxford is a registered trade mark of Oxford University Press in the UK and in certain other countries

First Edition published in 2017

Impression: 1

Published in the United States of America by Oxford University Press
198 Madison Avenue, New York, NY 10016, United States of America

British Library Cataloguing in Publication Data
Data available

Library of Congress Control Number: 2016952215

ISBN 978–0–19–878595–8

Printed in Great Britain by
Clays Ltd, St Ives plc

Für Mama und Papa

Contents

Preface

This is a book about counterfactuals. It is also a book about probability insofar as issues about probability shed light on our understanding of counterfactuals. As its main topic, the book studies the behaviour of counterfactuals in contexts of uncertainty. The guiding thought is that contexts of uncertainty provide a rich source of evidence which may help in deciding some of the most central issues in the debate about counterfactuals.

I first came into contact with conditionals when writing a term paper about Dorothy Edgington's 'On Conditionals' in Oxford in 2006. Dorothy later agreed to supervise my BPhil thesis on the problem of embedding indicative conditionals. I continued to work on conditionals in my dissertation, which I wrote as a member of the *Phlox* research group led by Benjamin Schnieder in Berlin, being supervised jointly by Benjamin and Dorothy. This book is a reworked version of my PhD thesis, submitted at the Humboldt University of Berlin in 2010. First motivated by my own desire to make some changes, then by extensive comments I received from readers at OUP, the manuscript has been revised in three stages, spreading over my time at the *Logos* group in Barcelona (2011–13) and my stay at the University of Tübingen (2013–16).

Many people helped with the project. First of all, I would like to thank my former supervisors Dorothy Edgington and Benjamin Schnieder for their long-lasting support over the years. Thanks are also due to the other members of the *Phlox* group, Nick Haverkamp, Miguel Hoeltje, and Alexander Steinberg. I am grateful to Sven Rosenkranz and Thomas Sattig for advising me on the book project during my time in Barcelona and Tübingen. Various people have provided me with helpful comments about material in the book, either in personal communication or at workshops and conferences. I clearly remember the following (apologies to those who have slipped my mind): Frank Arntzenius, Andrew Bacon, Matthew Bird, Jennifer Carr, Daniel Dohrn, Christian Folde, Wolfgang Freitag, Ulrich Gähde, Geert Keil, Angelika Kratzer, Thomas Kroedel, Maria Kuper, Hannes Leitgeb, Dan López de Sa, Ofra Magidor, Christian Nimtz, Bryan Pickel, Graham Priest, Tobias Rosefeldt,

Hans Rott, Robert Schwartzkopff, Wolfgang Schwarz, Wolfgang Spohn, Stephan Torre, Barbara Vetter, Lee Walters, Robbie Williams, Timothy Williamson, Alistar Wilson, Richard Woodward, and Elia Zardini. In this context, I also want to mention Katrina Przyjemski, a friend from my time in Oxford, who died in 2014. Various versions of material in this book have been presented at workshops and conferences in Padua (2007), St Andrews (2007), Hamburg (2007, 2014), Oxford (2007), Tilburg (2009), Berlin (2009, 2013), Barcelona (2011), and Konstanz (2013). I thank the organizers for giving me the opportunity and all the participants (particularly those I forgot to mention above) for their comments and suggestions. Special thanks go to Alan Hájek and Antony Eagle who provided, as readers for OUP, extremely detailed comments on earlier versions of the manuscript. I also thank two anonymous referees for their suggestions and the team around Peter Momtchiloff at OUP for their time and patience.

There are some friends who made life a lot easier and more fun at various stages of the project: Vera and Giuli, Giulia and Giovanni, Sophia and Thomas, and, as always, Katharina, Alex, and Kerstin.

I am happy to respond to comments at moritz.schulz[at]uni-hamburg.de.

Hamburg, Germany, 2017

An earlier version of chapter 6 was published as 'Counterfactuals and Arbitrariness', *Mind* 123, 1021–55. OUP kindly granted permission to reprint the material.

At certain stages of preparing the manuscript, I have benefited from partial funds by the Spanish Ministry of Science and Innovation for the project CONSOLIDER-INGENIO 2010 CSD2009-00056 on *Philosophy of Perspectival Thoughts and Facts (PERSP)* and for the project FFI2009-13436, I+D+i programme on *Semantic Content and Context Dependence.* I have also profited from the generous support of the *Deutsche Forschungsgemeinschaft* while working on the first draft of this book.

1

Introduction

A good deal of our reasoning—both practical and theoretical—takes place under premises which are uncertain. We are frequently not in a position to know all the relevant facts. When it comes to making a decision, a time constraint often forces us to make up our mind before all the information is in. In other cases, a piece of information may simply be very difficult to get or not available at all. Although reasoning from known premises may constitute an ideal we are aiming at, in practice we are often left with less to go on.

The situation is no different and probably even a little more acute for the counterfactual elements of our reasoning. It is very hard to know what would have happened if things had gone a different way—or, what matters most for our practical concerns, what would happen if we were to take a particular course of action. In many cases, our information will only suffice for making judgments with various degrees of likelihood.

This is a study about uncertain counterfactuals. The various degrees of likelihood we attach to counterfactuals form a puzzling pattern which challenges their standard semantics. As we shall see, the challenge might even run so deep that the very possibility of associating counterfactuals with truth conditions is called into question. Naturally, we would want to know whether the challenge can be met and what would be required for doing so. Here we can look to the debate about indicative conditionals for guidance, where a similar problem has been studied extensively. As far as counterfactuals are concerned, many questions are still open, which was the reason I thought the present study would be worth doing.

In this introductory chapter, I will present the main aspects of the phenomenon we will be concerned with, show how it leads to an interesting problem and sketch some general ways out of the difficulty, which will be discussed in more detail later. This will give an outline of what is going to come. A hasty reader already familiar with the problem and possible solutions to it could jump directly to a brief overview included at the end of this introduction.

1.1 Uncertainty about Counterfactuals

Why should uncertainty about counterfactuals be special in any way? This question is actually harder than it may seem, for answering it would ultimately turn on whether counterfactuals express ordinary propositions or perform some other function. In the first case, uncertainty about counterfactuals would not really go beyond what we are already familiar with in other areas of discourse. It would be of the same general kind as being uncertain about where my glasses are or what the weather will be like tomorrow. In the second case, however, uncertainty about counterfactuals might be special. It might, for instance, be a special kind of counterfactual uncertainty irreducible to uncertainty about a certain proposition. It is actually this tight connection with the question of whether counterfactuals express propositions which makes the study of uncertain counterfactuals so exciting.

In this section, I will present some preliminary considerations concerning potential sources of uncertainty about counterfactuals. Which factors can make us uncertain about a counterfactual? What I am going to say is only intended to get a general idea of why one may take counterfactuals to prompt unique ways of being uncertain about them. We will later discuss in more detail whether we should accept the relevant data as genuine or reject them as deceptive.

Uncertainty about counterfactuals can have, of course, a very ordinary source: the lack of knowledge about matters of fact on which the truth of the counterfactual depends. In a situation of fire, I may not know the height of the building I am standing on and the texture of the ground surrounding it. For this reason, I may not know whether I would survive if I were to jump off the roof. Nevertheless, on many ways of resolving my uncertainty about the building's height and its surroundings, I would be able to know whether the counterfactual is true. For instance, coming to know that the building is not very high and surrounded by soft ground is likely put me in position to know that I would survive. On the other hand, if it turned out that the conditions were rather different—the building being high, say, standing on a ground of concrete—I could probably know that I would not survive. A simpler example with the same structure might be provided by a situation in which I am about to play Russian roulette. My uncertainty about whether I would survive if I were to pull the trigger would mainly stem from my uncertainty about the location of the bullet in the cylinder.

Sometimes, however, it seems that our uncertainty about a counterfactual has more to do with the counterfactual itself rather than with ignorance about particular matters of fact. Going back to the fire example, we could equally well suppose that the ground is quite irregular: some stretches are soft while others are very firm. Moreover, the building might be of intermediate height, just in the middle of that zone where some people survive while others do not. Let us further suppose that I know all the relevant facts. So, I know the exact constitution of the ground, the exact height of the building and the bits of physics and medicine relevant to the scenario. Would I be in a position to know whether I would survive if I were to jump? Intuitively, it feels like we should say 'No': I might survive but I might equally not depending on exactly where and how I would land on the ground.

The situation is puzzling given that we assumed knowledge of all the relevant facts. If this is really so, can't we simply calculate the position I would land on together with the angle and velocity with which I would hit the ground? Combined with my physical constitution, this should give us knowledge about whether I would survive. Yet the problem is that in order to make this calculation, we would need to know where and how exactly I would jump off. And this does not seem to be completely determined by where I currently stand on the building. Almost every detail could matter: the speed and direction I would have when leaving the roof, the relative position of my body parts, etc. Clearly, some of these issues are informed by what we already know. For instance, there will be an upper limit on the speed with which I could possibly leave the roof and the way people like me tend to behave in situations like these will narrow things down a little further. But nothing seems to determine the exact way in which I would jump off the roof if I were to jump at all (assuming here that as a matter of fact I will refrain from jumping). In more theoretical terms: sometimes, there can be many relevant ways in which the antecedent of a counterfactual could be realized. If these different ways do not yield a uniform verdict about the truth of the consequent, then, intuitively, we are inclined to be uncertain about the counterfactual.

There is potentially a further source of uncertainty about counterfactuals. Sometimes it can also be uncertain how the world would have unfolded from the time onwards, after the antecedent had been made true. Consider the case of a lottery. Could we know whether we would have won if we had bought a ticket? Intuitively, the answer seems again to be 'No'. What we are inclined to say is something like 'Sure, our chances would have been slim, so we would probably have lost, but with some luck, we could have won.' Why would we qualify our counterfactual judgements in situations like this?

An analogy suggests itself. It seems that our thinking about possible counterfactual futures is interestingly similar to our thinking about actually possible futures (we come back to this thought in section 7.2). Just as we may be uncertain about how the ticket we have bought will perform in the lottery, we can be uncertain about how a ticket would have performed if we had bought it. What seems to change is the perspective from which we look at the possible futures: in one case it is our present position and in the other case it appears to be a counterfactual standpoint determined by the antecedent of the counterfactual.

As a tentative picture—to be further refined and clarified in what is going to come—we find that there are three potential sources of uncertainty for counterfactuals. To begin with, there is the very ordinary possibility of being uncertain about matters of fact on which the truth-value of the counterfactual depends. What is more interesting and, as we shall see, much more controversial, is that there seem to be two further sources of uncertainty deriving from the nature of counterfactual thought itself: (i) we can be uncertain about the exact way the antecedent would have been realized in the relevant counterfactual scenarios and (ii) we can also be uncertain about how a counterfactual scenario would have unfolded after the antecedent had been made true in one way or other.[1] The two sources have in common that they both introduce ways in which more than one counterfactual scenario could be relevant for the truth-value of a given counterfactual. This common consequence will be what we are mostly going to focus on.

I hope this provides an initial grip on how counterfactuals could stand in an interesting relation to uncertainty. What we need to see now is where

[1] Cf. the discussion of the roles of determinism and indeterminism in Edgington (2008).

we can look for data on uncertain counterfactuals. This is the topic of the next section.

1.2 Where Uncertainty Shows Up

Quite generally, uncertainty can show up at a number of different levels. Perhaps most fundamentally, it is captured by our *credences* or *degrees of belief* describing how likely we take a given proposition to be. Credences are belief-like attitudes which grade the credibility of a proposition on a scale, standardly represented in terms of the unit interval $[0,1]$. Epistemologically, a credence can be taken to reflect the degree of justification a subject takes itself to possess for the proposition under consideration.[2] As degrees of justification behave to a large extent like probabilities—a degree of justification representing how likely the proposition is made by the evidence—a perfectly coherent set of credences is often assumed to obey the basic axioms of probability theory.

We can test for intuitive credence assignments by asking various questions. How strongly do you believe that *p*? How certain/confident are you about *p*? Can you exclude that not *p*? There would also be more theoretically involved ways of testing for credences, but we will not make use of them, for they are hard to exploit in the case of counterfactuals.[3]

An important source of evidence comes from the expression of uncertainty in language. This source will be very important for us, not only because we can worry less about the objects of credences, but also because the data seem to impose stronger constraints on possible explanations. Uncertainty gets expressed in language in various ways. First of all, there are expressions of epistemic modality such as 'perhaps, possibly, maybe' or 'likely, probably'. But there are also explicit epistemic attitude ascriptions which ascribe graded epistemic attitudes. Consider, for instance, 'is

[2] There would be a whole lot more to be said. As credences are not identified with degrees of justification but are construed as the subject's best take on the latter, one would like to know more about how to understand a subject's *take* in this context. In addition, there are important aspects of justification not yet captured by this picture. Intuitively, there is a difference between a credence of 0.5 based on complete ignorance and a credence of 0.5 based on known chances, say (for further discussion, see Joyce 2005).

[3] These concern the interaction of credences and preferences in decision making, where credences may reveal themselves by favouring certain decisions over others when combined with a set of preferences. For general discussion, see Jeffrey (1965) and Joyce (1999).

fairly confident/certain' or 'considers it possible'. Embedding conditionals under such expressions promises to give us further evidence on uncertain conditionals.

To illustrate the different ways in which uncertainty can be detected, consider the lottery case again. We may assign a high credence to the counterfactual

(1) If we were to purchase a ticket, we would not win.

We can further test this judgement by embedding the counterfactual under epistemic modals and explicit epistemic attitude ascriptions expressing high probability:

(2) *It is probable/likely* that if we were to purchase a ticket, we would not win.

(3) We are *fairly confident* that if we were to purchase a ticket, we would not win.

Given that explicit epistemic attitude ascriptions and epistemic modals reflect our epistemic states, the former in a straightforward way, the latter in a possibly much more complex way, one would expect that there is a certain harmony between our credences and their expression in language. As a default, we may suppose that a high credence in a proposition is correlated with a probability modal being acceptable and an ascription of high confidence being fine. In application to counterfactuals, we could then suppose that intuitive credence assignments to counterfactuals correspond to the acceptability of embedding counterfactuals under an epistemic phrase expressing the relevant degree of belief. This is only a default, though, for it is at least conceivable that counterfactuals behave differently in embedded contexts from how they behave in unembedded ones. If that were the case, our credence assignment to what a counterfactual expresses in an unembedded context might depart from what we would say about a corresponding embedding of the counterfactual.

1.3 The Debate about Indicative Conditionals

For those familiar with the debate, it will by now be obvious that we are looking at a problem structurally similar to the problem concerning the relation between indicative conditionals and (subjective) conditional probabilities. Adams (1965, 1975) and Stalnaker (1970) conjectured—hinted at already by Ramsey (1929: 155, fn. 1)—that the subjective probability of a conditional should be the corresponding conditional probability.[4] To evaluate 'If *A*, *B*' would be to ask how likely it is that *B* is true conditional on the assumption that *A* is true. Call this *Adams's thesis*.[5] (A drawback of this label is that Adams explicates the thesis in terms of the *assertability value* of conditionals and not in terms of their credence. This is not what I shall mean by it.) There is strong evidence for something like this to be correct. Start by considering unembedded indicative conditionals like

(4) If the coin is tossed, it will come up heads.

(5) If I buy a lottery ticket, I will lose.

It is quite natural to take the second conditional to be highly likely and the first one to still deserve a credence of around 1/2. This pattern extends to epistemic embeddings of conditionals both under explicit epistemic attitude ascriptions and under epistemic modals:

(6) Mary considers it possible that if the coin is tossed, it will come up heads.

(7) John is fairly confident that if he buys a lottery ticket, he will lose.

(8) Perhaps, if the coin is tossed, it will come up heads.

(9) Probably, if I buy a lottery ticket, I will lose.

In such cases, the intuitive credence assignments and the embedding behaviour under epistemic expressions are in perfect correspondence with the conditional credences of the consequent given the antecedent. Based on such data, one may suspect that evaluating a conditional amounts to evaluating the consequent under the assumption of the

[4] It is not so clear what Ramsey actually thought about the relation between conditionals and conditional probability. In the vicinity of the frequently cited footnote, Ramsey (1929: 154–6), he also considers material and strict interpretations as sometimes adequate and in Ramsey (1926: 76) he explicitly denies that the probability of 'If *p*, *q*' is the corresponding conditional probability.

[5] This thesis and its variants go under many names: for obvious reasons, it is sometimes also called *the Ramsey test* or *Stalnaker's hypothesis*, but other names like *The Thesis* (Edgington 1995) and various acronyms exist for it, too. Admittedly, my choice is largely arbitrary, though I think that Adams deserves credit for first having given the thesis a rigorous defence as part of a fully developed theory of conditionals.

antecedent, which is a way to uncover the conditional probability of the consequent given the antecedent.[6]

Let us make Adams's thesis a little more precise. I will represent an indicative conditional with the double arrow '$\Rightarrow$'. Let us call an indicative conditional *simple* if the double arrow neither occurs in its antecedent nor in its consequent. For reasons which go beyond the present scope of our discussion, the evaluation constraint on indicative conditionals is usually restricted to simple conditionals.[7] Adams's thesis can then be given as follows:

Adams's Thesis. Let P be a subjective probability function and $\ulcorner A \Rightarrow B \urcorner$ a simple conditional. Suppose $P(A) > 0$. Then $P(\ulcorner A \Rightarrow B \urcorner) = P(B|A)$.

The conditional probability of B under A is only defined according to the standard definition if the probability of A is positive (see also section 2.1). This is why the proviso that $P(A)$ be greater than 0 is present. Adams's thesis is silent about how to evaluate indicative conditionals with an antecedent which is taken to be certainly false.[8]

Adams's thesis can be seen as imposing a coherency constraint on our subjective probabilities. It states how our credence in an indicative conditional should cohere with our credences in the antecedent and the conjunction of antecedent and consequent, for the conditional probability equals, if defined, the fraction of these two probabilities, that is

$$P(B|A) = \frac{P(A \wedge B)}{P(A)}.$$

Note further that it is a normative thesis concerning what the probabilities of conditionals should be. It can be expected that epistemic subjects will

[6] In certain contexts, indicative conditionals have a reading which does not conform to Adams's thesis (though an Adams friendly reading is usually available, too, or can at least be made salient by a slight change in context or perspective). See Kaufmann 2004 and McGee 2000 for examples.

[7] For how a proponent of Adams's thesis might deal with nested conditionals, see Edgington (1995: sec. 7.1).

[8] Adams (1996: 2f.) takes it to be the default assumption to set $P(\ulcorner A \Rightarrow B \urcorner) = 1$ if $P(A) = 0$. Schulz (2010b) contains some considerations why this might be a fruitful assumption. But see also Bennett (2003: §23) for the view that the probability of a conditional should really be undefined if the antecedent is assigned zero probability.

sometimes violate it. Yet in doing so they will be incoherent if Adams's thesis is correct.

Given that it is not clear whether Adams's thesis can be combined with a truth conditional semantics for indicative conditionals, I have presented it as applying to sentences, not propositions (see also section 2.1). So, the thesis can be given a minimal interpretation according to which indicative conditionals are associated with conditional probabilities in some way or other leaving the exact nature of this association open. This makes room for a wide spectrum of possible explanations.

In a number of cases, our credences in indicative conditionals can go counterfactual. We may have a positive credence in a coin toss counterfactual like

(10) If the coin had been tossed, it would have landed heads.

And as mentioned before, we intuitively take it to be likely that

(11) We would have lost if we had bought a lottery ticket.

As in the indicative case, the intuitive credence assignments are preserved under embeddings:

(12) Mary considers it possible that if the coin had been tossed, it would have come up heads.

(13) John is fairly confident that if he had bought a lottery ticket, he would have won.

(14) Perhaps, if the coin had been tossed, it would have come up heads.

(15) Probably, if we had purchased a lottery ticket, we would have lost.

Within a certain range of cases, the credence assignments to counterfactuals pattern in a way similar to how they play out for indicative conditionals. This is evidence that something like Adams's thesis should also be true of counterfactuals.

Despite its intuitive support, Adams's thesis is not unproblematic. Prominent semantic theories of indicative conditionals like the material analysis cannot be used to explain this thesis and are, when coupled with certain background assumptions, even incompatible with it. Given this mélange of intuitive plausibility and theoretical difficulty, it is not much of a surprise that Adams's thesis and its problems have prompted an incredibly fruitful debate resulting in many important innovations in the theory of conditionals. Variants of it have been discussed in different branches of philosophy, as well as in linguistics and to a some extent in psychology. Not all of these sub-debates were connected at the time. For instance, the work by Kratzer (see the collection of papers in Kratzer 2012), which became something like the standard view of conditionals in linguistics, was not fully taken up within the more traditional philosophical debate until fairly recently (as might be inferred from the fact that it is not mentioned in the extensive 2003 survey by Bennett; an exception is Lycan 2001).

Although there were some early suggestions that counterfactuals might display an interesting relation to conditional probability, too (Adams 1975, Skyrms 1981), the debate about counterfactuals was mostly occupied with a different set of problems. I can only speculate about why this was so. It may not have been clear that there is a similar but distinct problem for counterfactuals. To some it may have appeared to be basically the same problem in a slightly different guise. For those who do not see a significant semantic gap between indicative conditionals and counterfactuals, this stance would have been a natural consequence of their view. It could also have been a factor that there were independent problems with indicative conditionals (e.g. the paradoxes of the material analysis) which had a natural explanation in terms of Adams's thesis. In contrast, the theories of counterfactuals developed by Lewis (1973, 1979) and Stalnaker (1968, 1975)—the latter actually intended as a general theory of conditionals—were so elegant and powerful and provided such an illuminating description of the puzzling logical behaviour of counterfactuals that the basic problem of counterfactuals might have appeared to be solved, leaving only questions of detail open. Not until recently has it been observed by Edgington (2008) and Moss (2013) that counterfactuals display an equally challenging relation to conditional probabilities.

The debate about Adams's thesis and its variants has led to a rich solution space for the problem. Given the similarity between the data for indicative conditionals and counterfactuals, it seems natural to explore the problems for counterfactuals with an eye on the existing solutions for indicative conditionals. As will become evident in the course of our discussion, copying existing solutions one-to-one won't give us what we need. But we can take the debate about indicative conditionals as a guideline for asking various questions about counterfactuals and focussing on certain types of approach in our attempt to answer them. This captures to a large extent the research programme pursued in this study. It goes without saying that aiming even at rudimentary completeness would have been too ambitious an aim. So, I have concentrated only on a few basic questions. In a nutshell, these are: how can the data for counterfactuals be systematized? To what extent do they challenge semantic theories of counterfactuals? Is there a non-truth-conditional way of explaining the epistemic profile of counterfactuals along the lines of the restrictor view? Can we modify standard semantics in some way to achieve a truth-conditional explanation of the problematic data? In the remainder of this introductory exposition, I will briefly present the chapters addressing these questions and indicate how they can be seen as deriving from certain trends in the debate about indicative conditionals. To set the stage, we will have to start with some preliminary remarks on counterfactuals and their standard semantics.

1.4 Counterfactuals

To begin with, a few terminological points. I have decided to employ the label 'counterfactual'. Possibly under the influence of Lewis's book *Counterfactuals*, it seems to be the most common expression. Despite its popularity, it is widely held to be an obvious misnomer. The point is that 'counterfactual' is read 'counter-to-fact', which in turn is taken to suggest that counterfactuals imply or presuppose the falsity of their antecedent.[9] Let me stress that I use the label without any presumption in this direction. The cases discussed by Anderson (1951) and Stalnaker (1975: 145f.) provide good evidence that counterfactuals do neither semantically nor conventionally imply the falsity of their antecedent. And perhaps one does not have to hear 'counterfactual' as meaning that counterfactuals somehow imply the falsity of the antecedent. Perhaps one can hear this word to indicate that counterfactuals flourish when the antecedent is false. They are our best means of talking about unrealized possibilities.

9 Chisholm (1946) suggested this term as a label for counterfactuals, though he admitted that it is not quite adequate.

Counterfactuals have some grammatical features worth highlighting.[10] A striking property of counterfactuals is the forward time shift in the interpretation of tense in their antecedent (cf. Dudman 1983, 1984 and Gibbard 1981). In a free-standing sentence, a tense like past perfect is typically used to refer to a point in time which lies earlier than a salient point in the past. But past perfect does not receive this kind of doubly past interpretation in the antecedent of a counterfactual. There it is typically interpreted like simple past, but can also refer to the present or the future:

(16) If Jacky had observed the bank more closely, she would not have chosen it as a target.

(17) If Paul had been with Jacky in the bank right now, he would have told her to give up.

(18) If Jacky had reached the flight to Paris tomorrow, she would have used her falsified French passport. (context: she already got caught by the police.)[11]

A similar forward time-shift can be observed for counterfactuals with simple past in the antecedent. In the environment of the antecedent of a counterfactual, simple past gets interpreted as being about the present or the future. This phenomenon might be intimately connected with how counterfactuals generate the specific meaning they have, for it opens up the possibility that one dimension of tense is not interpreted temporally but receives an interpretation along a different dimension (for discussion, see Iatridou 2000).

As far as the consequent of counterfactuals is concerned, it is worth noting that it is usually governed by a past tense modal auxiliary as in the following example:

(19) If the police had followed her, Jacky *would/might/could/should* have noticed.

The modals occurring in the consequent are plausibly taken to be the past tense forms of 'will', 'may', 'can', and 'shall' respectively (cf. Dudman 1991 and Iatridou 2000). The fact that 'would' is likely to be a past tense form

[10] For an extensive survey on the syntax of conditionals in general, see Bhatt and Pancheva (2006). A detailed discussion of the grammar of counterfactuals can be found in Iatridou (2000).

[11] The last two examples are modelled after similar examples by Dudman (1984: 149f.).

of 'will' is going to concern us again in section 7.2. A quick final point: in general, I use 'counterfactual' to refer to 'would'-counterfactuals, but when the contrast with other counterfactuals like 'might'-counterfactuals matters, I try to be more explicit.

1.5 Standard Semantics

Largely due to the work of Lewis and Stalnaker, the debate about counterfactuals has produced a very elegant and powerful semantic framework for describing an extensive range of conditionals. In this study, I will make an attempt to explicate most issues in terms of the conceptual tools provided by it. I hope this will allow for easy comparisons with alternative proposals. So, to set the stage, let me briefly sketch the standard semantics of counterfactuals.

There is wide agreement that evaluating a counterfactual somehow involves taking into consideration possible situations in which the antecedent of the counterfactual is true. The semantic status of the counterfactual will then depend—in some way or other—on whether the consequent of the counterfactual is true in those situations. A view like this is famously expressed by Lewis. He opens his book *Counterfactuals* with the following statement:

> '*If kangaroos had no tails, they would topple over*' seems to mean something like this: in any possible state of affairs in which kangaroos have no tails, and which resembles our actual state of affairs as much as kangaroos having no tails permits it to, the kangaroos topple over. (Lewis 1973: 1)

According to this idea, the truth-value of a counterfactual depends on what is the case in similar possible worlds at which the antecedent of the counterfactual is true. When wondering whether kangaroos would topple over if they had no tails, we would have to consider the most similar worlds at which kangaroos have no tails and ask whether kangaroos topple over in these worlds.

In what follows, I will adopt a somewhat more abstract and less committal terminology by simply speaking of the *relevant worlds* at which the antecedent is true. So, I follow the standard approach in assuming that there is always a certain set of worlds in terms of which the truth conditions of a counterfactual can be specified. But speaking only of relevant worlds leaves open the question in which relation these worlds stand to the actual world. It may be that the pertinent notion of relevance can be cashed out in terms of similarity in certain respects, but I prefer to leave this as a substantial question which is not already decided by terminological choices. Also, the present terminology does not presuppose that the worlds which matter for the evaluation of a counterfactual are minimal according to a certain ordering by being, for example, the *most similar* or *closest* worlds. For all the terminology presupposes, a relevant world at which the antecedent is true does not have to be the *most relevant* world (cf. the discussion about the limit assumption in section 7.4).

Assuming that the truth conditions of a counterfactual depend on what is true at certain relevant worlds at which the antecedent holds, what exactly is the connection between the set of relevant worlds and the truth or falsity of the counterfactual in question? According to what can be called the *standard view*, there is a simple answer to this question: a counterfactual is true just in case the consequent is true at *all* relevant worlds at which the antecedent is true. For instance, it is true that kangaroos would topple over if they had no tails just in case *all* relevant worlds at which kangaroos have no tails are worlds at which they topple over.

Let us represent 'would'-counterfactuals by the corner '>'.[12] Then the standard semantics can be captured by the following clause:

> **Standard Semantics.** Let ϕ and ψ be sentences. A counterfactual $\ulcorner \phi > \psi \urcorner$ is true at a world w iff ψ is true at all relevant ϕ-worlds with respect to w.

Note that the notion of relevance is relativized to a world of evaluation w: with respect to different worlds, different sets of worlds may be the set of relevant antecedent-worlds.[13] For example, with respect to a world at which a given lottery is rigged, the set of relevant worlds at which I buy a lottery ticket may contain no world at which I win; however, with respect to a world at which the lottery is fair, the set of relevant worlds may well contain such a world.

[12] I prefer using the corner over the more common box-arrow '□→' because the box-arrow suggests that 'would'-counterfactuals involve an element of necessity, an assumption which I will ultimately reject (see the discussion of the duality thesis in section 7.2).

[13] For short, I will say 'antecedent-world' instead of 'world at which the antecedent is true'. Similar expressions such as 'consequent-world' are used similarly.

According to the standard view, counterfactuals behave very much like a universal quantifier on the set of relevant antecedent-worlds. The antecedent-clause is taken to select a set of relevant antecedent-worlds. The whole counterfactual is then assumed to be true just in case the consequent is true at all of those worlds. Thus, counterfactuals would express a kind of metaphysical necessity restricted to the set of relevant antecedent-worlds.

There are two important variants of the standard view, one being a slight deviation, the other being a special case. In his official semantics, Lewis departs somewhat from the standard view as characterized by his opening statement (see Lewis 1973: 19ff.). He does not assume that there is always a set of relevant antecedent-worlds, for he construes this set as the set of most similar antecedent-worlds which is empty if there is a non-terminating chain of more and more similar antecedent-worlds. We will discuss this worry in section 7.4.

Now to the special case. In his original semantics, Stalnaker (1968) assumes that there is always a single most relevant or closest antecedent-world. This can be seen as a special case of the standard view if the set of relevant worlds is assumed to always be a singleton set. However, I will use the expression 'standard semantics' or 'standard account' to be associated with the general presumption that there are usually many relevant antecedent-worlds. So, Stalnaker's original proposal will not be classified as a standard account. As it stands, Stalnaker's account is often perceived as problematic, for it seems to make too strong a metaphysical assumption by excluding the possibility that there can be two worlds which are equally close to the actual world.[14] Still, there are other phenomena in the light of which Stalnaker's semantics appears much more favourable (see e.g. the sections 7.2 and 7.3) and there are ways to soften this seemingly problematic consequence. As a matter of fact, the semantics I ultimately favour can be seen as a variant of Stalnaker's semantics with some elements of standard semantics integrated into it (see chapter 6).

With this in mind, let us get back to the behaviour of uncertain counterfactuals. Before we take a look at how data about uncertain counterfactuals might challenge standard semantics, let us first see how these data might be systematized.

[14] Subsequently, Stalnaker has refined his view in the light of such objections. See Stalnaker (1978).

1.6 Systematizing the Data

If we take the data about uncertain counterfactuals seriously, a good starting point would be to try and see whether we can describe them in some general way. Chapters 2 and 3 look at attempts to do this. As Adams's thesis answers the parallel question for indicative conditionals, we will take it as a role model. To recall, it states that the subjective probability of an indicative conditional is the conditional probability of the consequent given the antecedent. The question thus becomes whether a similar hypothesis can be formulated about counterfactuals.

Before we start reaching out for new conjectures, we should quickly check that it is not simply the same hypothesis we are looking for. Could Adams's thesis be plausibly applied to counterfactuals? If counterfactuals were very much like indicative conditionals, this could be expected. Conversely, if counterfactuals significantly differ from indicative conditionals, we might suspect that there are counterexamples to Adams's thesis when applied to counterfactuals in those cases in which the truth conditions of the two kinds of conditional come apart.

It has become an almost uncontested fact that a counterfactual can be false while the corresponding indicative conditional is true. The classic example which shows this is usually attributed to Adams (1970):[15]

(20) If Oswald had not killed Kennedy, someone else would have.

(21) If Oswald did not kill Kennedy, someone else did.

Clearly, the indicative conditional is true (or at least acceptable), for we know that someone or other killed Kennedy. But the corresponding counterfactual seems to be false. As far as we know, Oswald was not part of a conspiracy. He acted on his own. So, if Oswald had not killed Kennedy, no one else would have.

Once it is clear that there is a semantic difference between indicative conditionals and counterfactuals, we should be sceptical about the application of Adams's thesis to counterfactuals. This suspicion is confirmed

[15] As a matter of fact, Adams (1970) gives the example with the consequents negated intending to show that a counterfactual can be justified while the corresponding indicative conditional is not. This way of highlighting the difference between counterfactuals and indicative conditionals is more controversial (although it seems to come with a similar amount of intuitive support), for a counterfactual would then no longer imply the corresponding indicative conditional.

when we embed the Oswald/Kennedy conditionals under epistemic operators. Contrast the two sample sentences in an embedding under an expression of high probability:

(22) Probably, if Oswald did not kill Kennedy, someone else did.

(23) Probably, if Oswald had not killed Kennedy, someone else would have.

The first probability ascription is acceptable while the second is not. Given Adams's thesis, we can explain the plausibility of the first ascription by pointing to the fact that our present conditional probability of someone else having killed Kennedy under the assumption that Kennedy did not do it is high (it might even be 1). Given that this is so, the implausibility of the second ascription contradicts the idea that Adams's thesis might hold for counterfactuals, too. If it were to hold of counterfactuals, we should expect the probability of the counterfactual to be high, for our present conditional probability is high. But it is not. Hence, our credences in counterfactuals do not always correspond to our present conditional probabilities.

Not all may be lost, though. Perhaps Adams's thesis needs only to be amended slightly when extended to counterfactuals. The idea might be that counterfactuals trigger a shift in the perspective from which the conditional probabilities are evaluated. Something like this was indeed Adams's idea (see Adams 1975: ch. 4): he takes the present subjective probability of a counterfactual to be our *past conditional probability* of the consequent given the antecedent. So, in the Oswald/Kennedy case, the relevant conditional probabilities would be those we had a short while before Kennedy was killed. As it stands, this proposal cannot be quite correct. Think, for instance, of situations in which we now think that our past credences were not quite adequate. We would not want our present credences in counterfactuals to inherit our past mistakes about corresponding indicative conditionals.

Edgington (2008) abstracts from the details of Adams's proposal. She suggests that the probabilities of counterfactuals can be seen as conditional probabilities formed in a corresponding *hypothetical belief state*, which has no essential connection with any past belief state. For instance, the hypothetical belief state relevant for the Oswald/Kennedy case will include all the information we currently have about what had happened up to a short while before Kennedy was killed. It will not, however, contain the information that Kennedy was actually killed. In such a belief state, our conditional probability of Kennedy being killed given that Oswald is not going to do it can then be expected to be low. To complete this account, we would need, of course, a story about how hypothetical belief states can be derived from present belief states.

A possible answer to this question might be found in a suggestion by Skyrms (1981). He puts forward the idea that in evaluating counterfactuals we estimate the *(possibly past) conditional objective chances* of the consequent given the antecedent. For instance, why is it that we should think it likely that we would have lost had we bought a lottery ticket? Answer: Because our chances of winning conditional on playing were low. An important question about Skyrms's suggestion will concern the notion of objective chance it employs. What exactly is its nature? To put the question somewhat more constructively: how do we have to conceive of the notion of objective chance in order for Skyrms's proposal to be plausible as a fully general constraint on the evaluation of counterfactuals? This will be the leading question for chapter 3.

At this stage, it might be helpful to note that questions about the evaluation of counterfactuals are often similar to questions about the semantics of counterfactuals. For instance, the question of which hypothetical belief state is relevant for the evaluation of a given counterfactual is not very different from the question of which set of possible worlds is relevant for the truth conditions of the counterfactual. Now, given that we have a very powerful framework in which to address questions about the semantics of counterfactuals, it is a natural hope that we can make progress towards finding a precise evaluation constraint by making use of the tools provided by standard possible world semantics.

If we look at the data, it is striking that our credences in counterfactuals seem roughly proportional to how many of the relevant antecedent-worlds we take to be consequent-worlds. For instance, having a high credence in the counterfactual thought that we would have lost had we purchased a lottery ticket seems to come from realizing that most relevant possible scenarios lead to our ticket losing.[16] We may conjecture, then,

[16] It is not completely clear how a psychologically realistic picture of evaluating uncertain counterfactuals would look like. Playing out or mentally simulating certain scenarios in one's head and weighing them according to their respective likelihood can certainly play

that the kind of chance counterfactuals are related to is defined over the set of relevant antecedent-worlds as described in standard semantics. Work by Gärdenfors (1988), Joyce (1999), and Lewis (1976, 1981) on imaging functions develops a conception of counterfactual chance of this kind.

The upshot of chapter 3 will be—with some qualifications—the following evaluation constraint on counterfactuals:

(EC) The subjective probability of a counterfactual is an estimate of the (possibly weighted) fraction of consequent-worlds among the relevant antecedent-worlds.

Of course, this requires further explanation, but for the time being, it should suffice to give a first idea of what an evaluation constraint for counterfactuals could look like. Let us now see how it relates to standard semantics.

1.7 Identifying the Challenge

In the case of indicative conditionals, Adams's thesis has proven to be very problematic. Starting with Lewis (1976, 1986c), a number of so-called *triviality results* have been discovered showing that Adams's thesis can only be satisfied by a very limited number of probability functions. These results were accompanied by more direct arguments (Edgington 1986, 1995) to the effect that there is no conditional connective for which Adams's thesis can be assumed to hold. However, participants of the debate have drawn opposite conclusions from these results. Some, like Lewis, have rejected Adams's thesis (as construed here) and have looked for an alternative explanation of its intuitive plausibility.[17] Others, like Edgington and Adams himself, have taken the arguments to suggest that conditionals do not have ordinary truth conditions. The latter conclusion is possible because the problematic arguments not only assume Adams's thesis but also a propositional construal of the conditional: the probabilities of

a role, although it seems implausible that we always do so explicitly. A more direct grasp of the resulting structure seems often possible, which could be assumed to be shaped by our previous experience with similar counterfactuals.

17 The bracketed qualification is necessary because Lewis actually accepted what he took to be Adams's thesis, namely a probabilistic constraint on the assertability conditions of conditionals, which he added to the material analysis. However, he rejected Adams's thesis on its present construal, namely as a constraint on the credences in conditionals.

the conditional are taken to be probabilities attaching to a proposition expressed by the conditional.

Given that (EC), the evaluation constraint on counterfactuals, is structurally similar to Adams's thesis, we may expect to encounter similar problems with it. Due to Leitgeb (2012a) and Williams (2012), we know now that this is indeed the case. In chapter 4, I will mainly focus on developing a precise, but informally explicated puzzle with a similar upshot, based on an argument by Edgington (1995) for indicative conditionals. I hope this will make it easier to take into view possible ways of escaping this kind of problem. I come back to the triviality results in chapter 8.

To get an idea of how (EC) causes trouble, take a case in which most but not all relevant antecedent-worlds are consequent-worlds. A lottery might induce this structure: at almost all antecedent-worlds, my ticket loses, but in at least one of them, it wins (cf. Hawthorne 2005). Suppose further that the distribution of consequent-worlds among the antecedent-worlds is known to us. Then (EC) will prescribe a high credence in the corresponding counterfactual. This accords well with our intuitions about a lottery case, where it seems natural to be very confident that we would have lost if we had bought a ticket. However, we should have zero credence that the truth conditions prescribed by standard semantics obtain, for we know that at least one relevant antecedent-world is not a consequent-world and therefore that not all antecedent-worlds verify the consequent.

In the light of this observation, it may seem that the standard truth conditions are too strong. We should not reject a counterfactual already on the grounds that a few antecedent-worlds do not make the consequent true. So, we might weaken the standard truth conditions by requiring only that *most*, but not necessarily all antecedent-worlds are consequent-worlds (Bennett 2003: 249f. calls this the *near-miss proposal*). For a sufficiently large lottery, the lottery counterfactual will then come out true. But given that we can be taken to have the relevant semantic knowledge, we could then be certain that the counterfactual is true. Yet this over-predicts certainty: we cannot be certain that our ticket would have lost. For all we know, it might well have won (cf. Edgington 2008; more problems with Bennett's proposal are discussed in Hawthorne 2005).

It seems that the near-miss proposal assigns truth conditions which are too weak, whereas the standard account's truth conditions appear to be too strong. But it does not look like there are any possible truth conditions left in-between: if 'all' is too strong, but 'most' is already too weak even for arbitrarily high thresholds, how can counterfactuals have truth conditions? In rough outline, this is the puzzle the intuitive credence assignments to counterfactuals creates. In chapter 4, I will analyse this situation more closely. As it turns out, the argument is not entirely general. It leaves room for truth conditions of a certain kind if our epistemic access to counterfactuals can be assumed to be limited. This observation will be the basis for a possible solution in terms of a truth conditional semantics in chapter 6.

1.8 Non-Standard Explanations

If one is intrigued by the intuitive appeal of the evaluation constraint (EC) but discouraged by the trouble it causes for a semantics of counterfactuals, one might favour an approach which disentangles (EC) from the semantics. In other words, one would prefer an explanation of (EC) which is not based on the truth conditions of counterfactuals but explains it in some other way. Call an explanation of (EC) and, more generally, of the data giving rise to it *non-standard* iff the probabilities of counterfactuals and the corresponding probability ascriptions are not construed as probabilities of a proposition expressed by the counterfactual as a whole. On a *standard* explanation of a construction of the form ⌜Probably ϕ⌝—where ϕ might be a counterfactual—the truth conditions of the construction would go with how likely the proposition expressed by ϕ would be. In this section, I will point to various possible non-standard explanations and give my reasons for discussing in detail only one such kind in this study.

Perhaps the first strategy which springs to mind would to assume that the epistemic phrases have only superficially *wide scope* but are in fact interpreted as having *narrow scope*. What seems like an ascription of subjective probability to counterfactuals might in fact describe the subjective probabilities we would have had if the antecedent had been true. The problems (EC) causes would be due to a scope confusion.

To see that this cannot be the whole story, note that the following two attitude ascriptions are not equivalent:

(24) John is fairly confident that if he had bought a lottery ticket, he would have lost.

(25) If John had bought a lottery ticket, he would have been fairly confident that he would lose.

In fact, it may be that John would have been fairly confident that he would win if he had bought a ticket, for buying a ticket might always boost his confidence.[18] So, it does not seem that we can explain the data away by reconstructing wide scope ascriptions of probability on the level of surface structure as narrow scope ascriptions on the level of logical form.

A second approach might question that the data are really about the *subjective* probabilities of counterfactuals. Perhaps they concern a more objective kind of probability given that they seem to derive from a kind of objective conditional chance. There are two immediate problems with such a thought. First, the problems with (EC) are structural and have little to do with the kind of probability involved being subjective, which is why the problems could be expected to simply reappear (cf. Williams 2012). Secondly, such a proposal would not be plausible for explicit epistemic attitude ascriptions to counterfactuals. Epistemic attitude verbs simply do not allow for non-epistemic interpretations. However, some of the modal expressions such as 'probable' also have more objective interpretations. Here we have indeed to make sure that the examples work on the epistemic interpretation. As far as I can see, the epistemic interpretations are fine and even seem to be the more salient ones.

There is a further aspect of the evaluation constraint which stands in the way of finding easy alternative explanations in the vicinity of the two types of approach we have considered so far. This has to do with the fact that our subjective probabilities are described as an *estimate* of a certain kind of counterfactual chance. For this reason, we cannot take the probability ascriptions to be equivalent to explicit ascriptions of some kind of counterfactual chance, where the chance operator would govern only the consequent of the counterfactual. This becomes clearer when we consider a particular case. Suppose there is a coin. We know that it is likely to be a trick coin which always comes up heads (though this is not certain). In such a scenario, we can say

(26) It is quite likely that if the coin had been tossed, it would have come up heads.

What we should not assert, on the other hand, are any of the possible two ascriptions of objective chance:

[18] I owe this type of case to Timothy Williamson.

(27) If the coin had been tossed, it would have had an objective chance of 1/2 to come up heads.

(28) If the coin had been tossed, its objective chance of coming up heads would have been 1.

Our credence in the last two counterfactuals should be significantly less than 1, whereas the first ascription of high probability seems fine. The ascription of subjective probability in (26) cannot be taken to be equivalent to either of the two latter counterfactuals.

All this suggests that we need a more principled account of the data. Interestingly, the debate about indicative conditionals has produced two positions which can be seen as non-standard explanations of Adams's thesis: the *suppositional view* of conditionals and the *restrictor view*. According to the suppositional view as developed by Edgington (1986, 1995) based on Adams (1975), our credence in a conditional is not a categorical credence in a proposition expressed by the conditional as a whole, but rather a credence in the consequent of the conditional assigned within the scope of supposing the antecedent (see also Barnett 2006, Bennett 2003 and Gibbard 1981). The restrictor view—originating from Lewis (1975) and further generalized by Kratzer (1981, 2012)—can be seen as a more linguistic variant of this idea. According to it, 'if'-clauses are a means of restricting various operators and quantifiers. A probability ascription to a conditional is not seen as an ascription of probability to the whole conditional. Rather, the probability operator is taken to attach to the consequent only but relative to a new domain which has been up-dated with the information contained in the antecedent. To some extent, the restrictor view and the suppositional view can be seen as complementary, for the suppositional view is primarily concerned with explaining our credences in counterfactuals, whereas the restrictor view deals with the embedding behaviour of conditionals.[19]

It is very natural to ask whether something similar might work for counterfactuals. In our discussion, we will primarily focus on the

[19] A difference concerns speech-acts performed with conditionals: on the suppositional view, an affirmative utterance of a conditional is a hybrid speech-act consisting in supposing the antecedent and asserting the consequent under this supposition. The restrictor view can account for assertions of conditionals as ordinary assertions of a proposition (see chapter 5).

restrictor view and keep the suppositional view in the background.[20] And as far as the restrictor view is concerned, we will mainly deal with its *static* variant rather than its implementation into a *dynamic* semantic framework. This will spare us a lot of technicalities without much loss, for the issues we will be concerned with play out similarly in the two frameworks.

As it turns out, the restrictor view cannot be straightforwardly applied to counterfactuals: mere domain restriction is not sufficient for explaining the data. What we will need is rather a certain kind of *domain modification* which describes how the domain of an epistemic phrase can be modified by a counterfactual 'if'-clause to yield something like a hypothetical belief state from the perspective of which the consequent can then be evaluated (cf. the conception of a hypothetical belief state in Edgington 2008 as mentioned earlier). How all this might work will be the topic of chapter 5.

I have set aside a few further possibilities of explaining the data in a non-standard way. First, one might try to develop some kind of *pragmatic* account of why the data appear to be so persuasive. A possible role model could be Jackson's (1979) pragmatic explanation of Adams's thesis in terms of a special rule of assertion for indicative conditionals which connects them with conditional probabilities. It is straightforward to construct an analogue of this for counterfactuals: simply substitute the evaluation constraint for Adams's thesis in Jackson's account.[21] My reason for not following-up on this idea is that a Jackson-type approach does not provide us with an explanation of how embeddings of conditionals under epistemic phrases are supposed to work. Moreover, it is silent on how to account for our credences in conditionals (cf. Edgington 1995: 247). In order to complete the theory, one would probably have to adopt elements of the suppositional view or the restrictor view (for a sketch of this idea, see section 9.2). Hence, it seems to suffice to focus on the latter theories only.

There is a second type of approach which will not quite get the attention it might deserve, namely an error theory about the data, possibly

[20] For applications of the suppositional view to counterfactuals, see Barnett (2010) and Edgington (2008).

[21] See the recent work by Leitgeb (2012a,b). His theory is very rich and much more complex than the idea I have just sketched, but it contains the idea of isolating a pragmatic meaning of counterfactuals which is suppositional and not based on truth conditions. For criticism of Leitgeb's theory, see Hájek (2014).

accompanied with some pragmatic elements to soften its consequences (cf. Hájek 2007). Given how challenging the data are, an error theory might not be unreasonable. My reason for setting this possibility aside has to do with the fact that I see error theories as something like a last resort. We should only consider them when the chances of finding a positive explanation start to look dim. Given that I thought the data allow for a positive explanation, I did not see the need to give a possible error theory a lot of space. I will, however, come back to this possibility in the concluding chapter.

1.9 Modifying the Semantics

As we shall see, giving a non-standard explanation of the problematic data along the lines of the restrictor view is not without problems. It is not clear whether there is any independent evidence for the kind of domain modification needed to account for the interaction of counterfactual 'if'-clauses with epistemic phrases (see section 5.6). In addition, it turns out that a restrictor type view faces a certain problem of embeddings which suggests that counterfactuals under epistemic operators behave more like sentences expressing ordinary propositions (see section 5.7). The upshot of this is that it would be nice to have a standard explanation of the data in place, if only as a plan B in case the problems with the modifier story resist a satisfying solution.

In the chapter on how challenging the data are for propositional accounts of counterfactuals (ch. 4), we will see that the space of options is greatly reduced but a small niche is still left open. To develop a theory for that niche, we will have to ask whether our epistemic access to counterfactual facts might be limited in a principled way. If our credence of 1/2 in a counterfactual like 'If the coin had been tossed, it would have landed heads' comes from us not having any way of knowing how the coin would have landed, we may be able to escape the threatening counterarguments, for they tend to assume that such knowledge would, on a propositional construal of counterfactuals, always be possible. Intuitively, imposing limits on possible counterfactual knowledge is very much in line with the amount of counterfactual knowledge we are willing to credit ourselves with. For instance, in most circumstances we would naturally deny that we know whether the coin would have landed heads if it had been tossed.

The big question is, however, what a semantics for counterfactuals could possibly look like which is capable of predicting in a systematic way the desired limits on our epistemic access to counterfactual facts.

Taking a look at the corresponding problem for indicative conditionals, there are a few propositional explanations of Adams's thesis available (see e.g. van Fraassen 1976, McGee 1989 and, more recently, Bacon 2015). It is generally agreed that one can only underpin Adams's thesis with a propositional account of indicative conditionals by making indicative conditionals highly context-sensitive to the epistemic state from the perspective of which they are evaluated. Given that counterfactuals are standardly perceived to express more objective states of affairs, the subjective element in those theories is probably not a desirable feature of a theory of counterfactuals. But there is a second thought which can be derived not only from those proposals, but also from some many-valued approaches to indicative conditionals like the random-variable approach in Jeffrey and Stalnaker (1994): the idea of arbitrary selection. Fixing the conditional probability of *B* given *A* is very similar to asking how likely an arbitrarily selected *A*-world would be a *B*-world. For instance, when we ask how likely the coin is to come up heads given that it is tossed, we may as well ask how likely an arbitrarily selected toss-world is a heads-world. Can an idea like that be applied to counterfactuals? And what tools would we need to build a semantic theory around it?

The proposal I will make is to implement this general idea with the help of the epsilon-operator, originally introduced by Hilbert (1922, 1923) (see also Ackermann 1924). It allows one to denote arbitrary elements of certain classes. Just like a definite description operator, it forms a singular term out of a formula, but it does not require uniqueness: for the term to have a denotation, the relevant class does not have to contain only one element. The idea would then be to tie the truth conditions of counterfactuals to what is true at an arbitrarily selected relevant antecedent-world. In rough outline, the suggestion will be this:

(TC) A counterfactual $\ulcorner A > B \urcorner$ is true iff the *A*-world arbitrarily selected out of the set of relevant *A*-worlds is a *B*-world.

This defines Stalnaker-type truth conditions for counterfactuals: the truth-value of a counterfactual is construed as depending on what is the case at a single world. At the same time, however, it is granted that there are usually many antecedent-worlds relevant for the evaluation of a given counterfactual. In this way, the proposal circumvents the potentially worrisome observation that there might often be various equally relevant antecedent-worlds.

It goes without saying that the use of the epsilon-operator in specifying a semantics for counterfactuals raises many issues concerning, for example, a possible indeterminacy in the truth conditions of counterfactuals. On the other hand, it also gives us a rich tableau of explanatory possibilities. For instance, we may get into a position to explain the puzzling epistemic profile of counterfactuals by viewing it as coming from a counterfactual fact being represented as the result of an arbitrary selection of a world out of a number of relevant worlds. It is evident that this will impose limits on our epistemic access to counterfactuals: in general, the use of an epsilon-term provides no clue as to which of the candidate objects it denotes.

1.10 Overview

We will begin with an attempt to systematize the puzzling relation between counterfactuals and subjective probability. The two chapters 2 and 3 are about principles which unify the data about how counterfactuals tend to be evaluated, so that more informed consequences can be drawn from them (chapter 2 is more critical, while chapter 3 develops a positive account). The role model for such a principle will be Adams's thesis that the subjective probability of an indicative conditional should be the corresponding conditional probability of the consequent given the antecedent. As indicated earlier, my suggestion will be to follow-up on a proposal by Skyrms and describe the evaluation of counterfactuals as an estimate of a specific kind of conditional objective chance.

In the debate about indicative conditionals, the big problem with Adams's thesis is that it does not seem to be compatible with a truth conditional semantics. Accordingly, chapter 4 analyses how the evaluation constraint developed in chapter 3 bears on the possibility of giving a truth-conditional semantics for counterfactuals. We will see how a basic argument to the effect that counterfactuals do not have truth conditions can be constructed. This argument leaves us with two basic options. First, the argument can be accepted to be sound: the evaluation of counterfactuals cannot be explained in terms of truth conditions. Instead, one may look for an alternative explanation of the evaluation constraint. Secondly, one may deny one of the major premises of the argument. Since they are concerned with the epistemic profile of counterfactuals, it may be possible to escape the conclusion of the argument by holding that our epistemic access to counterfactuals is limited in a principled way. The subsequent two chapters are grouped according to these two basic options.

The most prominent non-truth-conditional explanation of Adams's thesis is based on the idea that 'if'-clauses can act as restrictors. On this construal, the trouble with Adams's thesis reflects a misconception of the logical form of conditionals when they interact with certain operators. Rather than taking the operator to have a single argument place for a conditional proposition, the 'if'-clause is taken to first restrict the operator which then attaches to the consequent of the conditional only. In the light of this possibility, it is a natural question whether the restrictor view can be extended to counterfactuals in a similar way to yield an explanation of how counterfactuals interact with epistemic phrases. This is the topic of chapter 5. We will see that such an explanation can only be had if we generalize the idea of a restrictor by allowing for a kind of modification of operators by counterfactual 'if'-clauses which is not restriction. It will turn out, however, that the resulting view runs into problems when it comes to more complex compounds of counterfactuals. There it seems that we would like to be able to give a propositional explanation of embedded counterfactuals after all.

The main chapter, chapter 6, then presents a view which escapes the basic argument against truth conditions. According to this view, counterfactuals can be given a possible worlds semantics which stays fairly close to standard semantics. However, instead of accounting for counterfactuals with a universal quantifier ranging over a set of relevant worlds at which the antecedent is true, the alternative semantics employs the epsilon-operator. On this view, counterfactuals are construed as being about an arbitrary (relevant) world *w* at which the antecedent is true. It will be shown how the evaluation constraint can be explained on the basis of such a semantics. The study ends with a chapter on possible applications of the view and a discussion of whether it runs into problems with the triviality results. In closing, I consider the possibility of developing an error-cum-pragmatics theory as a back-up plan and briefly comment on the prospects of extending the view developed in chapter 6 to indicative conditionals.

2

The Problem of Evaluating Counterfactuals

We have already seen in the introductory chapter that subjective probabilities attach to counterfactuals in an interesting and challenging way. In the present chapter, we will take a closer look at possible patterns underlying the probability ascriptions to counterfactuals. As indicated in the introduction, the behaviour of counterfactuals in situations of uncertainty seems to be fairly systematic. This impression is reinforced if we look at the evaluation of counterfactuals in contexts in which the relevant objective chances are clear. Consider, for example, the following sequence of counterfactuals:

(1) If the die had been rolled, it would have landed on an even number.

(2) If the die had been rolled, it would not have landed on an even prime.

(3) If the die had been rolled, it would have landed on some number or other.

Suppose we take the die to be fair and everything else to be normal. Then it seems that we should have increasing credences in these counterfactuals. The first counterfactual seems to deserve a credence of 1/2, the second a credence of 5/6, and the last one a credence of 1. These credences correspond in an obvious way to conditional probabilities. For instance, the probability of a die landing on an even prime conditional on being rolled is 5/6. In the light of such data, we may conjecture that our credences in counterfactuals are somehow coordinated by certain conditional probabilities. The question is how this intuitive connection between counterfactuals and conditional probabilities is best spelled out.

In this chapter, we will go through three major proposals in the literature on this topic, which will gradually improve our picture of how counterfactuals connect with conditional probabilities. We will start with a suggestion by Adams (1975) and a more general view Edgington (2008) proposed. We will then look at a conjecture by Skyrms (1981), which promises to resolve some of the open questions concerning Edgington's view. Skyrms's proposal comes very close to the account I ultimately favour, but we will still spend some time in the next chapter implementing his suggestion in the framework of possible worlds semantics (and connecting it with the work on imaging functions). There are two reason for doing so. The first one is that there are some residual issues with Skyrms's idea which can be handled with much more ease in a more general framework. The second reason is that moving the evaluation constraint closer to possible worlds semantics will make for easier comparisons with various semantics for counterfactuals, which tend to be cast in this framework. It also makes it more perspicuous what a semantics for counterfactuals should be able to explain (or to explain away).

A remark on what this and the next chapter aim to establish and what is not in their scope. We are going to seek a principle which allows us to systematize the data on how counterfactuals get assigned subjective probabilities. The minimal presumption is that we can agree on there being an intuitive connection between counterfactuals and conditional probabilities of some kind, even if we disagree about what best explains this intuitive connection or even think that it should rather be explained away. The latter set of questions will not be addressed in this chapter, but in chapters 5 and 6 (and briefly commented on in the concluding chapter). So, in the following two chapters, we only try to identify what about the subjective probabilities of counterfactuals needs explaining, not how this explanatory task is best undertaken. For this reason, we will not yet focus on possible semantics for counterfactuals.

2.1 Probability

Before we get started, a few preliminary remarks on probability are in order. I will briefly explain the terminology I adopt and address some more general questions. The latter include issues having to do with what the objects of probability are, how to distinguish subjective probability in a descriptive sense from a corresponding normative concept and how subjective probabilities contrast with objective chances. We also have to say something about conditional probabilities. Finally, we will take a brief look at Lewis's Principal Principle.

Probabilities are naturally taken to attach to possibilities, which may in turn be construed as sets of possible worlds. Depending on how the objects of credences are best modelled, one may even prefer to take subjective probabilities to attach to more finely grained propositions. Considerations in favour of the latter choice would come from the familiar problems with taking sets of possible worlds as the objects of belief. For instance, we may believe the Pythagorean theorem without thereby believing Fermat's conjecture, despite the fact that both theorems may be necessary, in which case they would both correspond to the universal set of all possible worlds (for discussion, see Stalnaker 1984: ch. 4 and 5 and the reply in Field 1986). If we substitute in this problem belief with high credence, say, we get a problem for taking sets of possible worlds to be the objects of credences.

In the debate about conditionals, it is often desirable to avoid the assumption that conditionals are associated with sets of possible worlds or a more finely grained proposition, for such an association encapsulates the assumption that conditionals have ordinary truth conditions. In the context of our discussion, it would be preferable not to pre-decide the question whether counterfactuals have standard truth conditions by a choice of what the objects of probability are. For this reason, it would be helpful to work with alternative objects of probabilities, at least until we settle on a certain semantics for counterfactuals. Taking sentences instead of propositions is perhaps the most straightforward way out of this difficulty. A probability associated with a conditional in a context would then not presuppose that the conditional expresses a (classical) proposition. It could be that the probability is rather a different kind of probability, a conditional probability perhaps. Or it might be a probability attaching to a more complex representational object than a set of possible worlds. If these sentences turn out to express propositions, we can later redescribe the probabilities of sentences as the probabilities of the propositions they express. If not, a more complex story might be needed, which can still be expected, however, to involve the content of counterfactuals. Temporarily working with sentences as the objects of probability is taken here only as a heuristic method, it is not meant as an account of what probabilities fundamentally attach to.

The first kind of probability I wish to consider are *credences* or *degrees of belief*. They reflect how confident an epistemic subject is concerning the truth of a given proposition or sentence. For instance, I may be highly confident that my lottery ticket will lose, fairly confident that my favourite team will not go out of the competition at the qualifying stage, and indifferent whether the coin will land heads or tails. In the terminology of credences, these three cases would be described by saying that my credence that my ticket will lose is very high, that my credence that my favourite team will not go out early is fairly high, and that my credence that the coin will land heads/tails is about 1/2. Having credence x in a proposition should not be conflated with an explicit belief that the proposition in question is x-likely. Credences are construed as qualifying the strength of our epistemic attitude towards a given proposition and need not be accompanied by any explicit beliefs about probabilities (but, of course, they can be).

We should also note that credences or degrees of belief are *descriptive*: they describe doxastic states. Since doxastic states can display all sorts of idiosyncrasies and are rarely, if ever, fully rational, one cannot assume that credences are governed by the standard laws of probability. Since real people's doxastic states are often even inconsistent, it is highly doubtful whether credences conform to any systematic constraints at all.

Bayesian epistemology is a research project which is concerned with *subjective probability*. However, Bayesian epistemology is usually not understood as a psychological enterprise, which would aim at an empirical theory about how our degrees of belief actually evolve. Rather, Bayesian epistemology investigates *ideal* degrees of belief as a *perfectly rational being* would have them. This allows Bayesians to ignore the idiosyncrasies and limitations of actual belief distributions. The epistemological goal is to gain insight into which subjective probabilities we *should* (or are best off to) assign in certain situations by studying perfectly rational degrees of belief. In this sense, subjective probabilities as studied in Bayesian epistemology are *normative*.[1]

[1] Alternatively, one could study instead objective degrees of justification or *evidential probability*. See Williamson (1997, 1998) and Williamson (2000: ch. 9–10). To some extent, evidential probability is quite similar to what sometimes goes under the heading

It is a matter of dispute which level of idealization serves the Bayesian's epistemological goal best. To gain a sufficiently rich mathematical structure, it is usually supposed that perfectly rational degrees of belief conform to the standard axioms of probability. Against this assumption it is often objected that the underlying conception of perfect rationality gets stretched too much by supposing, in particular, that perfect rationality requires assigning subjective probability 1 to every logical theorem. A subjective Bayesian who makes this assumption may insist that *perfect* rationality does indeed require this kind of logical omniscience. But she then confronts the question of how insights about perfect rational beings so construed can be applied to actual epistemic subjects with limited epistemic capacities. In the context of our discussion, these types of issues are less pressing. We only need to presuppose that it makes sense to speak of the subjective probabilities a subject should or should not assign in a given epistemic situation.

I will make an attempt to use 'credences' and 'degrees of belief' in a descriptive sense and 'subjective probability' in the more normative sense the Bayesian attaches to it. When the difference really matters, I might still talk about the subjective probabilities we *should* assign in a given situation, even though the 'should' would not really be necessary as subjective probabilities carry already a normative force.

On the other side of the spectrum, there are non-epistemic kinds of probability. Perhaps the most prominent kind is *objective (physical) chance*. There are two important aspects of objective chances worth highlighting. First, objective chances vary with time (cf. Lewis 1980). *Before the campaign, the chances of the democratic candidate winning had been about fifty-fifty. When her opponent made a terrible mistake, her chances of winning were significantly raised. After the successful election, her chances of winning the selection became* 1 *and will remain there forever.* Hence, at different times, a proposition or sentence can have various different objective chances of being true.

A second aspect of objective chances is a possible connection with indeterminism. If there are at a given time at a certain world *w* non-trivial objective chances (i.e. chances different from 1 and 0), then we may suppose that the world *w* is indeterministic in the sense that its history

of 'objective Bayesianism'. Cf. Williamson (2011) and Williamson (2009a) (note: these latter two references are to Jon Williamson).

does not at every time determine its future. In this way, objective chances can be seen as tied up with the most fundamental laws of nature a world is governed by (cf. Schaffer 2007). Not everyone takes objective chances to be intertwined with indeterminism. There might be room for 'deterministic chance' applying to certain higher levels of the world and leaving a possible connection with indeterminism for a more specific kind of objective chance: chances applying to the most fundamental physical level (see e.g. Glynn 2010 for such a conception). I will mainly use 'objective chance' in this more fundamental sense, but as far as I can see, most of what we shall say would apply *mutatis mutandis* to a more liberal conception of chance. (In section 2.4.2 we shall see that the chances relevant for the evaluation of counterfactuals are most likely chances compatible with determinism.)

Before we discuss how subjective probabilities and objective chances might be related to each other, we should briefly take a look at the concept of *conditional probability*. Not only is Lewis's Principal Principle formulated in terms of conditional probabilities, but for obvious reasons, they will also figure prominently in what is going to come later.

The conditional probability of A under B can be thought of as the probability of A on the assumption that B is true. So, it is the probability of A which results from ignoring the cases in which B is false. To illustrate, suppose we consider a uniform probability distribution over a finite set of possible worlds. Then every world is assigned the same probability and the probability of a possibility is proportional to the number of worlds at which this possibility obtains. In such a situation, the conditional probability of A under B is simply the fraction of A-worlds among the B-worlds ($= \frac{|A \cap B|}{|B|}$).[2] So, the conditional probability of A under B is given by how much of the region in logical space at which B is true is occupied by A. Now, when we consider non-uniform distributions, we also need to consider the respective weights of the A-worlds among the B-worlds. For example, considering a fair coin, the conditional probability of it landing heads under the assumption that it is tossed is 1/2, the fraction of heads-worlds among the toss-worlds. But now suppose we consider a trick coin which lands heads about 2/3 of the time. In this case, the conditional probability of it landing heads on the assumption that it is tossed is 2/3,

[2] Given a set A, $|A|$ is its cardinality.

not 1/2. The heads-worlds among the toss-worlds have two times more weight than the tails-worlds.

Formally, the conditional probability of A under B, $P(A|B)$, can be defined in this way:

Conditional Probability. $P(A|B) := \frac{P(A \wedge B)}{P(B)}$ if $P(B) > 0$.

This definition can be given the following rationale. The numerator is the probability that A and B both obtain. This is the absolute probability of the A-worlds which are also B-worlds. Now, when we divide this probability by the probability of B, we compare the probability of the A-worlds which are B-worlds to the probability of the B-worlds and so get an expression of the fraction of A-worlds among the B-worlds weighted by the respective probabilities.

The ratio definition of conditional probability leaves the conditional probability of A under B undefined if B is assigned probability zero, for division by zero is not defined. There are various possibilities to deal with this issue. The first and obvious one is to accept the consequence that conditional probabilities are sometimes undefined. A second possibility is to look for a natural extension of the ratio definition covering the case in which the denominator is undefined. A possible convention is to stipulate that the conditional probability $P(A|B)$ is 1 if $P(B) = 0$.[3] To some extent, this stipulation is structurally similar to the stipulation that a conditional is vacuously true if its antecedent is impossible. Finally, it is possible to drop the ratio definition altogether and take conditional probabilities to be primitive (absolute probabilities can then be defined in terms of conditional probabilities).[4] For our purposes, we will not have to choose between these options.

With this in mind, let us now come to a possible connection between subjective probabilities and objective chances. It is a natural thought that subjective probabilities are constrained by objective chances: in some way or other, subjective probabilities should track objective chances. Suppose a fair coin is going to be tossed tomorrow. How likely should we think it to be true that it will come up heads? 1/2, of course. Why? Because its present objective chance of coming up heads is 1/2. On the other hand,

[3] This stipulation is sometimes made in the debate about conditionals. See footnote 8 in chapter 1.

[4] For a defence of this approach, see e.g. Hájek (2003).

matching the objective chances is not always desirable. To see this, let us change the previous example slightly by supposing that the coin toss took place already. As before, I take the past objective chance that the coin would come up heads to be 1/2. Given that I possess no further relevant evidence, it still seems mandatory to assign subjective probability 1/2 to the statement that the coin came up heads. But now assume further that I am told that the coin came up heads. This should raise my credence significantly, possibly to a value as high as 1. However, it should not alter my verdict concerning the past objective chances. So, this is a situation in which my credences should not follow my knowledge of the corresponding objective chances because I possess further relevant evidence. We can take from these preliminary considerations that a systematic description of how subjective probabilities relate to objective chances should steer a middle course between situations in which a match between subjective probabilities and known objective chances is rational and situations in which a divergence between the two seems to be preferable.

A way of making the connection between subjective probabilities and objective chances precise is given by Lewis's formulation of the Principal Principle.[5] The Principal Principle can be seen as a systematic constraint on how our thinking about objective chances should influence our subjective probabilities. Let me present the Principal Principle here by quoting Lewis:

The Principal Principle. Let C be any reasonable initial credence function. Let t be any time. Let x be any real number in the unit interval. Let X be the proposition that the chance, at time t, of A's holding equals x. Let E be any proposition compatible with X that is admissible at time t. Then

$$C(A|XE) = x. \qquad \text{(Lewis 1980: 270)}$$

To begin with, a couple of clarificatory remarks. First, XE is the conjunction of X and E. Secondly, Lewis states his view in terms of reasonable initial credence functions. These are supposed to represent reasonable or fully rational epistemic states which have not taken in any (empirical) evidence. Once an initial credence function is conditionalized on a body of evidence, it represents a rational belief state possessing this

[5] See Lewis (1980). Some possible refinements are discussed in Lewis (1994) and the postscript to Lewis (1980) in Harper et al. (1981). For a more general discussion of the big picture, see Edgington (2004b).

evidence. So, to reiterate, the Principal Principle is directly concerned only with initial credence functions representing the beginning of a rational epistemic life. By assuming that later epistemic stages come from earlier ones by conditionalizing on the aquired evidence, we can use the Principal Principle to also draw inferences about non-initial credence functions.

An important issue in clarifying the Principal Principle concerns the role played by evidence *admissible* at the given time t.[6] Let us first see that allowing admissible evidence E to enter the principle is better seen as a generalization than a restriction of the principle. If one were to omit E from the principle, one would still arrive at a valid (but weaker) principle. The simpler principle would say that conditional on a statement of objective chance, an initial credence function should match the chance value. To instantiate this with an example: under the assumption that the objective chance of a coin landing heads is 1/2 and before we have aquired any (empirical) evidence, we should think it 1/2-likely that the coin lands heads. As a matter of fact, the weaker principle is implied by the Principal Principle if we assume, plausibly, that the necessary proposition consisting of all possible worlds (and thus excluding none) is always admissible.

The generalization achieved by allowing admissible evidence takes care of cases in which what we assume over and above the objective chances does not bear on the target proposition. It allows, for instance, that with respect to a future coin toss, we may additionally make an assumption about what happens in causally unrelated parts of the world. Of course, Lewis could not have generalized the Principal Principle all the way: some evidence must be inadmissible. For instance, evidence which directly reveals the truth-value of the target proposition must be inadmissible (as long as the considered chance value is below 1). Under the twofold assumption that the chance of heads is 1/2 but that, as a matter of fact, the coin will come up heads, our credence in heads should be 1 and not 1/2, contrary to what the Principal Principle would suggest if all evidence were admissible.

It is a difficult question which evidence should count as admissible. To a first approximation, a piece of evidence can be taken to be admissible at t

[6] I am indebted to Alan Hájek for helping me a great deal to get clear about admissibility.

if it is solely concerned with the history up to *t* and with what the laws of nature are (including general laws of chance). Histories are usually taken to be admissible, for what happened before a point *t* at which the chances are taken cannot easily trump the information provided by the chances. If the statement under consideration is itself solely about the past and thus made true by the relevant history, it will have chance 1 of being true at *t* (the past is never chancy). Hence, the evidence provided by the history agrees with the evidence provided by the chances. Now, if the statement is about the future and we only possess information about the past and the present chances, it seems that our credences should go with the chances. The historical information may bear on what we think about the chances, say by indicating that a coin may be biased, but once the chances are fixed, it is hard to see how the historical information could trump the credences prescribed by the objective chances. For mixed cases of statements partly about the past and partly about the future, this line of thought does not go through as smoothly. And there are some hard cases, in which historical information may encode information about the future, say by suggesting that a time traveller testified that the coin will come up tails.[7]

Moving on, laws of nature and general facts about chance can be taken to be admissible for the following reason. Laws of nature do not seem to be chancy—they have always chance 1. The same can be held for general laws of chance, unless one opts for a Humean account of chance which allows for *undermining*: positive chances of sequences of events which would imply that the present chances are different from what they actually are. If this possibility is granted, the Principal Principle is in need of modification (see Hall 1994, Lewis 1994 and Thau 1994 for discussion and a possible way out). In connection with counterfactuals, we will have to say a little more on admissibility later (section 3.3).

Essentially, the Principal Principle then comes to this: conditional on the assumption that the objective chance of *A* at time *t* is *x* and conditional on the admissible evidence at *t*, one should (initially) have credence *x* in *A*. The principle can be seen as a specification of the idea that subjective probabilities are constrained by what we take the objective chances to be.

[7] See the discussion of 'crystal balls' in Hall (1994). For discussion relating the Principal Principle to possible knowledge of the future, see Hawthorne and Lasonen-Aarnio (2009) and the response by Williamson (2009b).

Prima facie, this principle is highly plausible, and it seems to allow for correct predictions with respect to a great variety of cases.[8]

2.2 Prior Epistemic Probabilities

A first step to associate counterfactuals with certain conditional probabilities was taken by Adams (see Adams 1975: ch. 4 and Adams 1976). He considers the idea that counterfactuals are, in a sense to be made precise, *epistemic past tense* forms of the corresponding indicative conditionals. More precisely, he suggests that the subjective probabilities of counterfactuals are tied to the prior epistemic probabilities of the corresponding forward-looking indicative conditionals, where forward-looking means that the conditional has a present tense antecedent and a future tense consequent formed with 'will'. To give an example, our present credence in the counterfactual

(4) If Oswald had not killed Kennedy, someone else would have

is taken to be coordinated with our past credence in the indicative conditional

(5) If Oswald does not kill Kennedy, someone else will.

Since Adams took the probability of an indicative conditional to be the conditional probability of the consequent given the antecedent, we would look at the thesis that the subjective probability of a counterfactual is the past conditional probability of the consequent given the antecedent. In this section, I will try to clarify Adams's proposal a little further and will also consider an interesting variant of it.

The basic idea of Adams's proposal is expressed in the following quotation:

> Whatever justifies the posterior affirmation of the counterfactual is then exactly the same as what did or might have justified the prior affirmation of the indicative, so the counterfactual can be regarded as a kind of *epistemic past tense*, expressing

[8] But see Hawthorne and Lasonen-Aarnio (2009). They mention counterexamples to the Principal Principle which stem from the realm of the contingent a priori. Cf. also the discussion in Schulz (2011). As a possible source of the problem, see the considerations in Williamson (2006) to the effect that contexts of subjective probability are hyper-intensional whereas contexts of objective chance are non-hyper-intensional.

not what *was* the case at a prior time, but what *could have been affirmed* at the prior time. (Adams 1975: 109)

In the quoted passage, it does not become clear how the phrases 'might have justified' or 'could have been affirmed' are to be understood. The immediate surroundings of the quotation suggest that these phrases are only meant to deal with cases in which we have not asserted or not even considered the corresponding indicative conditional at a relevant prior time. If we abstract from possible restrictions of this kind and assume that our credences are defined over all relevant sentences at the required times, we arrive at the thesis that we should think about a counterfactual now what we once thought about the corresponding indicative conditional.

That this proposal cannot be right is fairly obvious. Whatever the past epistemic state is in which the indicative conditional should be evaluated, we may gain information in the meantime relevant to the evaluation of the counterfactual. Someone may even tell us whether the counterfactual is true, or a crucial piece of evidence may become available which decides the truth-value of the counterfactual. In such cases, it is clear that our verdict about the counterfactual will depart greatly from our previous verdict about the corresponding indicative conditional. It would be very odd if what we presently think about a counterfactual were forced to inherit any mistakes we might previously have made about an indicative conditional.

To illustrate the problem, consider the following variant of the Oswald/Kennedy example. Suppose we had thought before Kennedy's assassination that there is a conspiracy plot to kill Kennedy. For this reason, we might have accepted the indicative conditional 'If Oswald does not kill Kennedy, someone else will'. Now time passes and we learn that there actually never has been any such conspiracy. As a result, we do not accept the corresponding counterfactual 'If Oswald had not killed Kennedy, someone else would have'. In such a scenario, our credence in the counterfactual does not go with our past credence in the forward-looking indicative conditional because the latter credence was based on a belief we now think is false.

Given this problem, one might wish to entertain an alternative interpretation, or rather a variant, of Adams's proposal. Instead of associating our present credences in counterfactuals with a possibly mistaken past credence in the corresponding indicative conditional, we may consider the idea that the right epistemic perspective from which to evaluate the indicative conditional is not our past epistemic state but rather our present epistemic state. Our credence in the counterfactual would then go by our present subjective probability concerning what was or would have been expressed by a past utterance of the indicative conditional. This seems to be a much more interesting proposal, for it would avoid the present problem and also seems to come with more initial plausibility. In order to properly evaluate it, we need to take a step back and contemplate the idea that counterfactuals might be past tense forms of the corresponding forward-looking indicative conditionals.

To be precise, let us consider the following view: (past perfect) counterfactuals presently express what a forward-looking indicative conditional expressed at an earlier time (cf. Dudman 1988, 1991, Strawson 1986 and Woods 1997: ch. 8). Call this *the past tense view* of counterfactuals.[9] For instance, the suggestion would be that the two Kennedy-conditionals above, (4) and (5), express the same content when the indicative conditional is uttered at an appropriate earlier time.[10] Basically, the past tense view says that counterfactuals and future-looking indicative conditionals differ merely in temporal reference. Evidence for this view mainly comes from the observation, described in the introduction (section 1.4), that 'would' can be seen as the past tense of 'will', which supports the idea that 'would'-conditionals might be past tense forms of 'will'-conditionals. Note, though, that this evidence is not fully conclusive, for other elements present in counterfactuals might interfere and tense in counterfactuals is, as we saw as well, not always interpreted temporally in counterfactuals.

Now, what has the past tense view to do with the variant of Adams's proposal we are presently envisaging? First of all, the past tense view implies

[9] In the literature, this view is sometimes referred to as the *relocation thesis*. Cf. Bennett (2003: 13).

[10] One may want to know what counts as an appropriate earlier time. A possible, though not fully specific answer would be that it is a point in time which lies shortly before the antecedent is made false in the actual world (assuming here that it is false). One may also doubt that such a point always exists—think of a counterfactual which concerns all times like 'If gravity obeyed an inverse cube law, then the planets would not have stable orbits' (thanks to Alan Hájek for this example). What is the appropriate earlier time? If one does not wish to border on inconsistency by saying that this point lies before the beginning of time (if such exists), one should perhaps say that in cases in which the counterfactual makes no specific temporal reference, there is no appropriate earlier time and the corresponding indicative conditional should be construed similarly as being about all times. It would then make no difference at which time it is evaluated.

Adams's proposal. If a counterfactual expresses what an indicative conditional expressed at an earlier time, then it seems clear that our present credence in a counterfactual should be our present credence concerning what is expressed by the earlier utterance of the indicative conditional, for these two credences would concern the same content. Conversely, if Adams's proposal were correct, this would be strong evidence for the past tense view. For suppose the past tense view were false. Then there would be a counterfactual and a corresponding forward-looking indicative conditional, where the latter would at no time express the same content. But if the expressed contents differ, we probably could have different credences concerning the two, in which case Adams's proposal would be false. Essentially, then, it seems that the past tense view describes the underlying semantic principle needed for the presently envisaged variant of Adams's proposal.

In order to evaluate the past tense view, let us tease out one of its consequences. Quite generally, if a sentence s_1 differs from a sentence s_2 merely in temporal reference and the tensed elements in s_1 refer in a context C_1 to a time t and the tensed elements in s_2 refer in a context C_2 to t as well, then an epistemic subject who is aware of the passage of time should have the same epistemic attitude towards s_1 in C_1 as she has towards s_2 in C_2 unless she gains or loses relevant information while transiting from C_1 to C_2.[11] Call this the *continuity principle* (cf. Schulz 2010a). For instance, if I am aware of the passage of time and neither gain nor lose any relevant information, I should think today about 'Today is my birthday' what I thought yesterday about 'Tomorrow is my birthday'. If our epistemic attitudes attach primarily to propositions, and the relevant pairs of indexical sentences are taken to express the same proposition in the two contexts respectively, then the continuity principle can be seen to follow from the more general fact that we should not change our beliefs unless we gain or lose relevant information. But since I do not want to preclude the view that our epistemic attitudes attach to propositions *as presented under the guise of a sentence* and want to leave open the possibility that the relevant pairs of sentences do not express quite the same proposition or no proposition at all, I prefer the somewhat more cautious continuity principle.

[11] I assume that the sentences are not context-sensitive in other respects.

If we apply the continuity principle to the past tense view of counterfactuals, we find the following consequence. An epistemic subject with a certain credence in a forward-looking indicative conditional at a given time should have the same credence in the corresponding counterfactual at some later point in time, provided she is aware of the passage of time and neither receives nor loses any relevant information. If a counterfactual merely expresses what an indicative conditional expressed at an earlier time, our earlier credence in the indicative conditional should equal our present credence in the counterfactual in situations in which we neither gain nor lose any relevant information. We can use this consequence to test the past tense view. Note that the problem cases for Adams's original proposal do not conflict with this consequence, for they were cases where we received information about the two conditionals (in the modified Kennedy case we learned that there actually was no conspiracy). The past tense view only predicts that our past credence in the indicative conditional should become our present credence in the counterfactual in those cases in which we do not receive any relevant information. This is a much weaker consequence, but as we shall see now, it is still too strong.

Against the past tense view, consider the following case.[12] Officer Shaker comes home in the afternoon. He finds a note in the kitchen saying 'I will be back tonight'. Unfortunately, he cannot identify who has written this. It may have been his daughter Linda or his wife Elizabeth. It is much more likely that it was written by Linda, because Elizabeth told him that she would be going on a business trip and would not be back before the other day. But Elizabeth's plans may have changed on short notice. In any case, Shaker seems right in thinking:

(6) If Linda is not home tonight, Elizabeth will be.

After this, Shaker receives a call from work and has to spend the whole night away from home. What should Shaker think at the dawn of the new day about the following counterfactual?

(7) If Linda had not been home last night, Elizabeth would have been.

Shaker has to consider two possibilities. Suppose first the most likely case that Linda wrote the letter. Then the counterfactual can be assumed to

[12] Bennett (2003: 346) discusses an example which is similar in many respects.

be false: no one would have been home last night if Linda had not been home last night, for Elizabeth would still have been on her business trip without any intentions of coming home early. In the second scenario where Elizabeth wrote the letter we could assume the counterfactual to be true. Still, since the first case is much more likely, Shaker should think that the counterfactual is probably false. Yet he took the indicative conditional to be highly likely or even certain. He does not seem to have received any relevant information. If the two conditionals differed merely in temporal reference, Shaker's different epistemic attitudes would not be rational. But they seem to be.

Here is a possible diagnosis. The counterexample works because there is merely an evidential link between the antecedent and the consequent of the indicative conditional. Since Shaker knows that the letter was written either by his daughter or his wife and he has no reason to distrust its content, he can conclude that if Linda is not home tonight, Elizabeth will be. But the available information does not support any non-evidential connection (or lack thereof) between antecedent and consequent.[13] As far as he knows, if Linda wrote the letter, Elizabeth has no intentions of coming home that night and her plans are independent of Linda's whereabouts. Hence, later on, he has not much reason to think that if Linda had not been home, Elizabeth would have been.

So, even the more plausible variant of Adams's proposal cannot be quite correct. Sometimes, what we think about a counterfactual is not what we think about an earlier utterance of the corresponding indicative conditional. We may think it unlikely that if Linda had not been home, Elizabeth would have been. Still, we may think that what the indicative conditional 'If Linda is not home, Elizabeth will be' expressed at an earlier time is quite likely. It seems, then, that even the modified proposal promotes too strong a link between counterfactuals and corresponding indicative conditionals. What we would need is an even more modest suggestion. Instead of past or present credences about past utterances, one may consider credences in a hypothetical belief state appropriately related to our present beliefs. To this possibility we shall now turn.

[13] The bracketed qualification is present because counterfactuals may occasionally owe their truth to the absence of any connection between antecedent and consequent. For instance, Elisabeth could have thought that she will come home independently of what Linda does and therefore hold 'If Linda did not come home, I would come home (and if she did, I would come home, too)' (thanks to Alan Hájek for this point).

2.3 Hypothetical Epistemic Probabilities

Adams (1975: ch. 4.8) already contemplated the possibility that his proposal may not hold in full generality. This led him to conjecture that our present subjective probabilities of counterfactuals may rather go with certain *hypothetical* subjective probabilities of the corresponding indicative conditionals, which may differ from our actual prior credences. Subsequently, this view was further developed by Edgington (1995, 2008).[14] Let us have a look at how she describes it:

> [T]he conditional probabilities relevant to the assessment of subjunctive conditionals do not (typically) represent your present actual distribution of belief, but those of a hypothetical belief state in a different context, normally that of an earlier time, concerning, for example, whether someone else will kill Kennedy if Oswald doesn't. (Edgington 2008: 8)

Edgington suggests that our epistemic attitude towards the counterfactual is determined by our epistemic attitude towards the forward-looking indicative conditional 'If Oswald does not kill Kennedy, someone else will' in some hypothetical belief state suitably related to our present belief state. Presumably, such a hypothetical belief state will include what we now believe about the circumstances a short while before Kennedy was killed, e.g., that Oswald prepared himself for the assassination. However, it will not include our present belief that Kennedy was killed by someone, for otherwise the forward-looking indicative conditional 'If Oswald does not kill Kennedy, someone else will' would be acceptable in the hypothetical belief state, which would yield the wrong verdict that the counterfactual 'If Oswald had not killed Kennedy, someone else would have' is acceptable, too. Let me make a few comments about this approach.

The hypothetical belief state approach is quite flexible and is not committed to the past tense view of counterfactuals, although it departs from it only slightly. It shares with the past tense view the more general assumption that the evaluation of counterfactuals is to be explained in terms of the evaluation of indicative conditionals, which are in turn assumed to go with the corresponding conditional subjective probabilities. This general assumption might still be disputable. A source of potential trouble is the

[14] Woods (1997: ch. 8) seems to favour a similar position.

contingent a priori.[15] Obvious a priori truths should not be denied in any belief state, or at least only under very specific circumstances. However, some a priori truths are contingent. Their contingency makes it possible that they can be truly denied in the consequent of a counterfactual. But then there are acceptable counterfactuals to which there are no natural hypothetical belief states in which the consequent is assigned a high conditional probability.

For instance, consider the following example:

(8) If there had been a massive space impact early in evolution, there would have been no thinkers.

This counterfactual seems to be acceptable (or could be made so by stipulating the relevant facts): if there had been a massive space impact early in evolution, conditions on earth would have been too unfriendly for higher animals to develop. However, in any belief state, the subjective probability that there is a thinker should be 1, for this is a fairly obvious a priori truth. If so, there is no reasonable hypothetical belief state in which the conditional probability of the consequent given the antecedent will be high: the conditional probability that there are no thinkers given that there was a massive space impact early in evolution will always be low. But then our present credence in the counterfactual does not correspond to the subjective probability of the consequent given the antecedent in a related hypothetical belief state. To reinforce the point, contrast the counterfactual with an indicative variant which tries to get at the same point:

(9) If there was a massive space impact early in evolution, there never were nor will be any thinkers.

As long as we do not doubt that we are thinkers, the indicative conditional is not acceptable. Even on the assumption of the antecedent, we would rather doubt that the space impact had the said effect.

Not much depends here on the choice of the example. All that is required is that there are some contingent propositions which should be assigned credence 1 in any belief state which is a candidate for modelling

[15] The following argument is inspired by considerations in Jackson (1987: 74f.), Weatherson (2001), and Williamson (2006).

the relevant hypothetical belief state. The latter qualification is important, for it shows that we do not even have to presuppose that the example is strictly a priori. For instance, one may think that we could doubt that there is a thinker by supposing that we are all robots of sorts which only execute programs but do not really think. The point would then be that in evaluating a counterfactual like the one above, nothing seems to invite us to consider such outlandish cases. A belief state in which we doubt that we are thinkers does not seem to be a good candidate to model the relevant hypothetical belief state.

Do the present considerations refute the hypothetical belief state approach conclusively? Probably not. What they seem to show is that the most straightforward interpretation of it cannot be quite correct. But there is still room for manoeuvre. Let me briefly mention one possible way out.

Perhaps *hypothetical* belief states are relevantly different from *ordinary* belief states. Such a difference may make it possible that obvious a priori truths can reasonably be denied in a hypothetical belief state. Two-dimensional semantics could provide the resources to implement this idea. In a two-dimensional framework, instances of the contingent a priori are associated with two propositions, one being necessary, the other contingent (cf. e.g. Chalmers 2006). For instance, a sentence such as 'There is a thinker' can be associated (in a given context) with a *primary proposition* which is necessary and a *secondary proposition* which is contingent. It seems possible to hold that our subjective probabilities are usually sensitive to the primary proposition expressed by a sentence. To continue with the example, the suggestion would be that we can be certain that there is a thinker because our credence attaches to the primary proposition expressed by 'There is a thinker', which is necessary. Now, this framework may be brought to bear on the present issue in the following way. Perhaps hypothetical belief states differ from ordinary belief states by being sensitive to the secondary proposition expressed by the relevant sentences. As a consequence, it would be possible to assign in a hypothetical belief state a low credence to 'There is a thinker' because hypothetical credences would attach to the secondary proposition expressed by the sentence under consideration. In the present example, the secondary proposition is contingent and differs from the primary proposition, which is necessary. On this picture, counterfactuals would invite us to think about the world as possibly non-actual, which may shift the perspective from which we evaluate the relevant sentences.[16]

At the present stage, it is not clear whether such a strategy will ultimately work. But it is likely that a fully developed account would (a) not be simple and (b) involve substantial theoretical commitments. Hence, it may not deliver an easily accessible principle which systematizes the data concerning uncertain counterfactuals on which further theorizing could then safely be based. Recall that we presently do not yet seek a theory of counterfactuals but merely a comprehensive way of capturing the relevant data so that we get a clear grip on what such a theory would need to explain.

2.4 Prior Propensities

The previous considerations can be taken to indicate that the subjective probabilities of counterfactuals do not stand in any systematic relation to the subjective probabilities of indicative conditionals. If one accepts this diagnosis, one may reckon that our credences in counterfactuals aim at something more objective. A natural way of fleshing this out would be to suppose that the evaluation of counterfactuals is guided by what we think about corresponding objective conditional chances. Skyrms (1981) has developed a view along such lines, albeit for somewhat different reasons. More recently, Moss (2013) has defended his approach. In this section, I will present Skyrms's proposal and point to some residual issues, which we shall resolve in the next chapter.

Skyrms introduces his suggestion as follows:

> My suggestion is that the prior probability account can be saved by one small change—the probabilities involved are the prior *propensities* rather than the prior epistemic probabilities. [...] [I]f we do not know for certain the values of the prior propensities, we may have to do with a weighted average—the expected prior propensities. The weights in this average will be epistemic probabilities, and we should use the best ones available—for this job—the *posterior* epistemic probabilities. I will use PR for epistemic probabilities and pr for propensities. I will superscript i and f for prior (initial) and posterior (final) respectively. Let the double arrow, $\Rightarrow$, symbolise the subjunctive conditional, and BAV be

[16] An approach of this kind could take further inspiration from the two-dimensional semantics for conditionals developed in Weatherson (2001).

'Basic Assertability Value'. Then the theory that I am suggesting can be succinctly expressed thusly:

*** $\text{BAV}(p \Rightarrow q) = \sum_j \text{PR}^f[\text{pr}_j^i] \cdot \text{pr}_j^i(q \text{ given } p)$.

where the pr_j^i's are the appropriate prior propensity distributions. (Skyrms 1981: 260f.)

Skyrms describes our present subjective probabilities of counterfactuals as aiming at the *prior objective conditional propensities* of the consequent given the antecedent. It is notoriously difficult to say what exactly a propensity is.[17] For the sake of simplicity, but also because I think it yields the most interesting account, I will tentatively equate propensities with objective chances (this assumption will be lifted in the next chapter). Thus, subjective probabilities of counterfactuals would be described as estimating certain prior conditional chances. To get a better grip on the proposal, let us apply this account to two examples. Consider the counterfactual

(10) If the coin had been tossed, it would have landed heads.

In Skyrms's notation, lower case letter pr^i represents the prior chances and capital letter PR^f our present subjective probabilities. Suppose we believe outright that the coin is fair. So, we think that its conditional chance of landing heads given that it was tossed was 0.5 (formally: $\text{pr}^i(q \text{ given } p) = 1/2$). In Skyrms's notation, this means $\text{PR}^f[\text{pr}^i] = 1$. Applying the formula, we find

(11) $\text{PR}^f[\text{pr}^i] \cdot \text{pr}^i(q \text{ given } p) = 1 \cdot 0.5 = 0.5$.

In the present scenario, we should have credence 0.5 in the counterfactual (10) because we are certain that the objective chance of it landing heads given that it was tossed was 1/2. This generalizes: if the relevant objective chances are known with certainty, the constraint recommends adjusting our credence in the counterfactual to what we take to be the past chance of the consequent conditional on the antecedent. For a great variety of basic cases, this seems to predict the intuitively correct values.

Let us also take a somewhat more complicated epistemic situation to illustrate the complexity of Skyrms's formula. Suppose we have credence

[17] Skyrms's account of propensities can be found in Skyrms (1980).

0.5 that the coin was fair. So, $PR^f[pr^i_1] = 0.5$ with $pr^i_1(q \text{ given } p) = 0.5$. Suppose also that we have credence 0.5 that it was biased towards heads with a probability of 0.7. In formal terms, $PR^f[pr^i_2] = 0.5$ with $pr^i_2(q \text{ given } p) = 0.7$. At this point, one crucial aspect of Skyrms's idea, namely that counterfactuals are evaluated by *estimating* prior conditional chances, shows up. Skyrms suggests that our credence in the counterfactual should now be raised to 0.6, since it should be the average of the relevant two chances weighted by our credences that these are the correct chances:

(12) $\sum_j PR^f[pr^i_j] \cdot pr^i_j(q \text{ given } p) = PR^f[pr^i_1] \cdot pr^i_1(q \text{ given } p)$
$+PR^f[pr^i_2] \cdot pr^i_2(q \text{ given } p) = 0.5 \cdot 0.5 + 0.5 \cdot 0.7$
$= 0.25 + 0.35 = 0.6.$

Indeed, although intuitions may not confirm the exact values, it seems correct that the more likely we take it that the coin was biased towards heads, and the stronger we take it to be biased, the more confidence we should have in the counterfactual thought that it would have landed heads if it had been tossed.

Before we turn to the details of Skyrms's account, let us see how it allows us to deal with the problematic issue surrounding the contingent a priori. The crucial fact to be observed here is that instances of the contingent a priori can have a chance to be false. It is a priori that there is a thinker, but at one time in the past, there may have been a chance that there would not be any. The contingency of instances of the contingent a priori makes it possible that they can have a chance to be false. So, if evaluating counterfactuals involves estimating *objective* conditional chances rather than hypothetical *subjective* probabilities, it will no longer be a problem that instances of the contingent a priori can be denied in the consequent of a counterfactual.

Skyrms states his view in terms of what he calls the *basic assertability value*. Thus, it is presupposed that assertability comes in degrees, i.e. that a statement may be more or less assertable (in this he follows Adams 1975, 1976). This is not an uncontentious assumption. It may well be that assertability is an all-or-nothing matter (cf. e.g. Williamson 1996). For instance, it would seem that the counterfactual 'If the coin had been tossed, it would have landed heads' is plainly unassertable, although it gets a basic assertability value of 1/2 on Skyrms's account (if the coin is taken to be fair). Given that we are looking for a characterization of the subjective probabilities of counterfactuals, I will interpret Skyrms's account in this way and talk about subjective probabilities instead of assertability values.

There are some residual issues with Skyrms's proposal. They mainly concern the exact nature of the notion of chance present in his account. In particular, we have to address Morgenbesser cases, the relation between chance and determinism and the possibility of counterfactuals with an antecedent contradicting the laws of nature. We also have to take a brief look at whether modus ponens holds for counterfactuals. As we shall see, each problem can be seen as providing information about how we have to understand the notion of objective chance for Skyrms's proposal to be plausible.

2.4.1 Morgenbesser Cases

To begin with, let us have a look at the original Morgenbesser case and a similar example discussed by Edgington (2004a):[18]

(13) If I had bet on heads, I would have won.

(14) If I had caught the flight, I would be in Paris now.

Initially, we may have credence 1/2 in (13) because we assume that the past chance of winning conditional on betting was 1/2. Similarly, we may be almost certain about (14) given that we take the past chance of being in Paris conditional on taking the flight to Paris to be very high. These credences are in accordance with Skyrms's constraint, for they are an estimate of the past conditional chances. However, they seem to be unstable. Suppose we find out that the coin came up heads. We can then be certain that I would have won, because how the coin would have landed would not have been influenced by my betting on the outcome.[19] An analogous situation can arise with respect to the second example as described by Edgington. Assume that we find out that the plane crashed because the on-board computer broke down. Given this discovery, we should have credence identical or close to 0 in (14), because if I had caught the flight, I would be dead now. Thus, our credences in counterfactuals

[18] The Morgenbesser example is reported in Slote (1978: 27, fn. 33).

[19] There may be trick coins which are somehow sensitive to my betting behaviour. If necessary, we could stipulate that we are dealing with an ordinary coin.

may be raised and lowered by evidence concerning events which occur after the time with respect to which the objective chances are evaluated.

Following a possible reaction considered by Slote (1978: 27, fn. 33), Phillips (2007) denies that the intuitions about Morgenbesser cases are genuine if the underlying chance process is truly indeterministic. I can see that this is a possible option (but cf. the defence in Edgington 2004a). However, if possible, we should prefer a framework which is not hostage to fortune in the sense that it would not be flexible enough to account for Morgenbesser cases if necessary. This is why I treat the Morgenbesser cases here as genuine. But nothing in the final theory will commit us to this position.

If genuine, Morgenbesser cases are a problem for Skyrms's proposal, for they would show that our credences in counterfactuals cannot always be our best estimate of the past conditional chances. Sometimes certain facts would be held constant in the evaluation of counterfactuals although they are made true only after the time from which the relevant counterfactual worlds are allowed to depart from the actual course of events. As a result, the past conditional chances can be knowably high or low without our subjective probabilities of the corresponding counterfactual being high or low respectively.

It is possible to amend Skyrms's proposal slightly to accommodate Morgenbesser cases. We could adopt a suggestion by Edgington about how to handle the problem:

> The objectively correct value to assign to such a counterfactual is not (or not always) the conditional chance of *C* given *A* at the time of the fork; but the conditional chance, at that time, of *C* given *A&S* where *S* is a conjunction of those facts concerning the time between antecedent and consequent which are (a) causally independent of the antecedent, and (b) affect the chance of the consequent. (Edgington 2004a: 21)

A brief clarificatory remark: by 'the time of the fork' Edgington means the point in time at which counterfactual worlds are allowed to depart from the actual history of the world in order to make the antecedent of the counterfactual true. This is the time at which we would expect the past chances to be taken. Now, the basic idea of Edgington's is to enrich the condition relative to which the conditional chances are evaluated. Instead of the chance of the consequent given the antecedent, we would work with the chance of the consequent given the antecedent plus a certain body of further information S. The sentence S will vary between different antecedents as it is concerned with information independent of the antecedent, and a sentence may be independent of one antecedent but dependent on another. For this reason, I think it would be more perspicuous to write S_A instead of S. This way it would be explicit that S is a function of the antecedent of the counterfactual.

In the examples above, S_A will comprise the information that the coin landed heads or the information that the plain crashed. That is to say, S_A will specify the information we hold constant in the evaluation of counterfactuals about times after the time at which the past chances are taken. If we incorporate this idea, we could rightly predict, for example, that we should have a low credence in (14): given that I took the flight plus the information that it crashed, the past chances of me being in Paris now would indeed be low.

Edgington also offers an account of the information contained in S_A. According to her, it is information about what is causally independent of the antecedent. This can be motivated by the observation that the reason why we sometimes hold certain information constant seems to be that the antecedent would not have had an influence on it. For instance, whether or not I had bet, it would not have affected how the coin would have landed. In a similar vein, one could also suppose that probabilistic independence from the antecedent at the relevant time is what matters in these cases. In addition, Edgington proposes a second condition saying that S_A should only contain information relevant to the probability of the consequent. This condition is probably there in order to limit the size of S_A. Without it, S_A could be blown up by a lot of irrelevant information.

Hájek (2014) raises doubts about Edgington's idea to identify S_A in terms of causal independence. However, in the context of our present discussion, the exact nature of such a set S_A is somewhat less important. Even if Edgington's specific proposal about what makes up the set S_A has to be amended, her general structural proposal to conditionalize on A together with a variable body of information S_A could still be correct. The set S_A might have to be defined in a slightly (or completely) different way, or it might not be definable at all. For now, it suffices to see that a proposal of this general kind could solve the problem with Morgenbesser cases for Skyrms, never mind whether Edgington's specific proposal is already fully satisfactory.

2.4.2 *Counterfactuals and Determinism*

Questions about determinism and indeterminism bear in many ways on the theory of counterfactuals. In the present context, our focus will be on whether the chances in Skyrms's evaluation constraint are chances which would require the world to be indeterministic. In order to answer this question, we will have to take a closer look at how the assumption of determinism influences the evaluation of counterfactuals. As we shall see, counterfactuals are not evaluated much differently under the assumption of determinism and so we will conclude that what we need is some kind of 'deterministic chance' which is largely indiscriminate between determinism and indeterminism.

For present purposes, the following rough-and-ready definition of determinism will do: a possible world w is *deterministic* iff for all times t, the history of w up to and including t together with the laws of nature obtaining in w completely determine what happens in w after t. So, if H_t is a complete expression of the history of w up to t (which is true at a world w' just in case w' shares its history with w up to t) and L is a complete expression of the laws of nature obtaining in w, then w is deterministic just in case for all times t the conjunction of H_t and L implies everything which in fact happens in w after t. In a deterministic world, there are at the fundamental level no non-trivial objective chances: at every time, the chances that a certain proposition is true will be either 1 or 0.

Now, how are counterfactuals evaluated on the assumption of determinism? *Prima facie*, it seems that this assumption does not make it much harder to evaluate counterfactuals and their evaluation does not seem to change a lot compared to indeterministic scenarios. In typical epistemic situations, we still think it about 1/2-likely that *if the coin had been tossed, it would have landed heads* regardless of whether we assume the lottery to be governed by deterministic or indeterministic laws.

To be sure, let us look at the issue in a little more detail. At first sight, one may think that if a world is deterministic, the set of relevant antecedent-worlds to be considered in the evaluation of counterfactuals consists either solely of worlds at which the consequent is true or solely of worlds at which the consequent is false. The reason for this assumption would be that in a deterministic world, the laws of nature together with the antecedent of the counterfactual either imply the consequent or its negation. If something like this were true, then one might wonder whether our uncertainty about a coin-toss counterfactual is simply grounded in our uncertainty about whether or not the antecedent-cum-laws imply the consequent. Our uncertainty would then not be informed by a certain conception of objective conditional probabilities which assume an intermediate value like 1/2.

As it stands, the present assumption cannot be correct, for the laws of nature together with the antecedent will in general not imply much, let alone a whole future which would always decide the consequent of the counterfactual. What is missing is information about initial conditions not yet specified by the antecedent. Perhaps the thought should rather be that the laws of nature together with the antecedent and a certain slice of the history of the actual world decide the consequent. Taking in a certain part of the history of the actual world is well motivated, for it seems that we typically hold a certain part of the past constant in the evaluation of counterfactuals. However, if the antecedent of the counterfactual is false, then any part of the history of the actual world together with the laws of nature is already inconsistent with the antecedent. In a deterministic world, a certain part of the history of the world together with the laws of nature already determines the whole future. If, by assumption, the antecedent is false, then its negation is implied by the history and the laws of nature. From this we can infer that the set of relevant worlds at which the antecedent is true cannot consist of worlds which share a certain history with the actual world and display the same laws of nature.

In the debate about counterfactuals, there are two main reactions to this issue. On the first proposal, the assumption that the relevant antecedent-worlds share a certain history with the actual world is given up (cf. Bennett 1984, but Bennett 2003: ch. 14 now defends the opposite view). In numerous possibly undetectable ways the history of the relevant worlds will at any time be slightly different from the history of the actual world. On the second proposal advertised by Lewis (1979), the assumption that the actual laws of nature obtain at the relevant antecedent-worlds is given up. To make the antecedent true, small 'miracles' are allowed to happen at the relevant antecedent-worlds. They are, however, only miracles relative to the actual laws. Relative to the worlds at which they occur, they are not miracles and compatible with the laws at those worlds, which will thus be different from the actual laws, however slightly.

Now, on both proposals, there is nothing in the concept of determinism which suggests either (a) in which way the history of the relevant antecedent-worlds should differ from the history of the actual world or (b) which kind of miracle should happen at the relevant antecedent-worlds. Since neither the difference in history nor the kind of miracle is uniquely determined, the set of relevant antecedent-worlds will often consist (a) of worlds which depart in slightly different ways from the history of the actual world (for various departures from the history of the actual world may be equally relevant) or (b) of worlds at which slightly different miracles happen (for various miracles may be equally relevant). Hence, there is no reason to suppose that the set of relevant antecedent-worlds will be homogenous in the sense of consisting solely of worlds which make the consequent true or solely of worlds which make it false.

To reiterate, consider a coin-toss case again and suppose that it is a deterministic process. As a matter of fact, the coin was not flipped. How would the world have been if it had been flipped? It might have been flipped in many different ways, it seems. Within certain boundaries, direction and momentum of the flip could have varied freely. Given the symmetries of the coin, the set of possible flips would most likely have produced a number of heads-ups close to 50%. For this reason, we may expect about 50% of the relevant toss-worlds to be heads-worlds even under the assumption of determinism.

If all this is correct, the chances in Skyrms's evaluation constraint are best construed as chances which can be non-trivial even in deterministic worlds. Skyrms's (1980: 24–6) conception of propensity was indeed 'deterministic' in this sense.

2.4.3 Counterlegals

There is a closely related issue. Counterfactuals may have an antecedent which contradicts the laws of nature (such counterfactuals are often referred to as *counterlegals*). A proposition incompatible with the laws of nature may never have a chance of being true. This seems clear for fundamental physical chances.[20] But even if we avail ourselves of a certain kind of non-fundamental chance compatible with determinism, there is no guarantee that it will assign positive probability to negations of laws of nature. What this would indicate is that the chances we need should not

[20] Maybe not. Humeans like Lewis (1994) think that chances can be undermined. Note, though, that undermining futures with a positive chance at some time do not seem to comprise all metaphysical possibilities. For this reason, deferring to a kind of 'Humean chance' would probably not solve our present problem.

only be compatible with determinism but also *global* in the sense of being well defined for all metaphysically possible antecedents.

How much of a problem are counterlegals? It seems clear that they are used less frequently than other counterfactuals. Still, counterlegals do not seem to be defective. What the world would be like if the gravitational constant had been different, say a little higher, seems to be a sensible question. So, an evaluation constraint on counterfactuals should accommodate the possibility that counterlegals may be justified with variable strength.

Some philosophers think that the laws of nature are metaphysically necessary.[21] However, if one thinks that the laws of nature are necessary, one will probably not think that counterfactuals whose antecedent is impossible are acceptable already on these grounds. In other words, they would not think that counterlegals are vacuously true, for this would probably make too many counterfactuals vacuously true. Hence, assuming the necessity of laws of nature would not free us from finding a conception of chance relevant to the evaluation of counterfactuals according to which laws of nature may have a chance of being false.

2.4.4 Probabilistic Modus Ponens

As noted by Adams (1976: 18), we seem to judge a counterfactual to be definitely false or incorrect if we take the antecedent to be true and the consequent to be false. In other words, modus ponens does not seem to be invalid for counterfactuals. In the context of a probabilistic evaluation constraint on counterfactuals, this leads to a problem, for the past conditional chances can be high even if the consequent turns out to be false despite the antecedent turning out to be true.

Here is the problem in some more detail. Sometimes the world unfolds in a way which was at an earlier time quite unlikely to happen. For instance, it may have been quite likely that I would lose if I were to buy a lottery ticket. But suppose I bought a lottery ticket and won against all odds. In such a situation, it seems we should not invest a high credence in the counterfactual

[21] See e.g. Edgington (2004b), who also relates this idea to some themes in Kripke (1980), or Smith (2001).

(15) If I had bought a lottery ticket, I would have lost.

Arguably, our credence should actually be zero.

If we take the evaluation of counterfactuals to be governed by an intuitive conception of past chances, then this is a problem. For what we think about the case is compatible with us taking the past conditional chances of losing to be high. It was probable back then that we would lose given that we were yet to buy a ticket. But on the present interpretation, Skyrms's proposal would recommend a high credence in the counterfactual even if we take the antecedent to be true and the consequent to be false. Thus, a probabilistic version of modus ponens would fail: a high credence in the antecedent combined with a high credence in the counterfactual would not generally warrant a high credence in the consequent. (If counterfactuals have ordinary truth conditions, it is not hard to see that a pattern like this would induce failures of modus ponens.) Note also that this would happen rather frequently and occur for all sorts of counterfactuals independently of any specific features of antecedent or consequent. Hence, they could not be assimilated to counterfactual versions of McGee's (1985) counterexamples to modus ponens, which employ a structurally complex consequent.

Skyrms (1980: 99f.) accepts this consequence and we should at least admit that intuitions may diverge on these cases because counterfactuals with an antecedent known to be true tend to be a little odd, most likely for pragmatic reasons. Although I take modus ponens to be (probabilistically) valid for counterfactuals, a weaker desideratum could, I think, be agreed upon by all parties: it would be nice to have a way of articulating an evaluation constraint on counterfactuals which is flexible enough so that we can have modus ponens if we want it.

2.5 Conclusion

We learned from the discussion of Adams's epistemic past tense view of counterfactuals that their evaluation should not be tied too closely to the evaluation of indicative conditionals. Counterfactuals do not express what indicative conditionals express either presently or at some past point in time. The discussion of Edgington's hypothetical belief state approach revealed that it might even be problematic to explain the evaluation of counterfactuals solely in terms of subjective probabilities. They seem to go by something more objective.

Skyrms's proposal solves some of these issues by describing the evaluation of counterfactuals as an estimate of the corresponding conditional objective chances. Some residual issues remain, however. It would be nice to get a clearer picture of the conception of objective chance needed to endorse the evaluation constraint in full generality. Our discussion has shown that the conception of chance involved should be largely indiscriminate between determinism and indeterminism and it should be global in the sense of being well defined for all metaphysically possible antecedents. Furthermore, it would be desirable if it were flexible enough to accommodate Morgenbesser cases and the validity of modus ponens.

We will see in the next chapter that a notion of *counterfactual chance* might be the best way of construing the concept of chance in Skyrms's constraint. We will also see that it is the same notion which has been used by Gärdenfors (1988) to generalize the notion of imaging as introduced by Lewis (1976). This will define an operation which ultimately gives us something like Edgington's hypothetical belief states. Modulo minor side conditions, the two ways of developing an evaluation constraint on counterfactuals in terms of counterfactual chance will turn out to be equivalent.

3

Counterfactual Chance

In order to further develop the notion of chance in Skyrms's proposal, it will be helpful to go back to the informal characterization of how counterfactuals are evaluated. When we ask how likely it is that *B* would have been true if *A* had been true, we seem to consider a number of relevant counterfactual scenarios in which the antecedent is true and estimate how many of them make the consequent true. This idea is already contained to some extent in Skyrms's approach, for oftentimes the objective chances it appeals to will target a set of close antecedent-worlds, but the idea seems to be somewhat more general in not putting any prior constraints on the nature of the relevant antecedent-worlds. We shall see that there are two approaches one can follow in developing the informal characterization further. Common to both approaches are two factors: a conception of relevance and a conception of chance which is designed to possibly weigh the relevant worlds.

The chapter is structured as follows. I start by discussing selection functions which can be used to describe the notion of relevance. Then a general notion of (metaphysical) chance is introduced which combines with relevance to provide a conception of counterfactual chance. With this in place, the notion of counterfactual chance can be used to formulate an evaluation constraint on counterfactuals which is structurally similar to Lewis's (1980) Principal Principle. After some further discussion, I will, in the second part of this chapter, take a look at Lewis's notion of imaging. The chapter closes with some more general considerations about the methodology used in introducing the concepts we need.

3.1 The Selection Function

To begin with, let me introduce the framework of selection functions as described in Lewis (1973: ch. 2.7). Based on an underlying set $\mathcal{W}$

of possible worlds, we may introduce a *selection function* f which takes us from a world $w \in W$ and a set of worlds $A \subseteq W$ to a set of possible worlds $f_w(A) \subseteq W$. The selection function is intended to model the notion of *relevance*: the set $f_w(A)$ can be thought of as the set of relevant A-worlds from the perspective of w. So, they are the worlds we should take into consideration when evaluating a counterfactual whose antecedent expresses a proposition true at exactly the worlds in A. For our purposes, we may actually identify propositions with sets of possible worlds.

There will be more to say on the notion of relevance in play (see section 3.9), but for now it suffices to merely note the structure of how relevance could be modelled in terms of a selection function which assigns to a world w and a proposition A a set of relevant worlds at which A is true. One could think of those worlds as being closest or most similar to w, but as noted in the introduction, this is not necessarily an assumption we have to make.

I should perhaps also say that I set aside a possible context-dependence of the notion of relevance. If counterfactuals are genuinely context-dependent, a natural place for accommodating this feature would be to let the selection function f be dependent on context, so that the selected set of worlds, $f_w(A)$, could vary between contexts despite the world of evaluation and the input proposition staying constant (cf. the discussion in Lewis 1973: sec. 3.1 and 4.2). The reason why I omit representing a possible context-dependence in the semantic model is that I think issues of context-dependence are largely orthogonal to the problem set I am trying to solve.

Further, let $[\![\bullet]\!]$ be an *interpretation function* which assigns to a sentence ϕ the set of worlds at which ϕ is true, that is $[\![\phi]\!] = \{w \in W : w \models \phi\}$.[1] The interpretation $[\![\phi]\!]$ of a sentence ϕ specifies the *truth conditions* of the sentence. According to this conception of truth conditions, two sentences have the same truth conditions if they are true at exactly the same possible worlds or, in other words, if they are necessarily equivalent. Now, given a sentence ϕ, the set $f_w([\![\phi]\!])$ can be thought of as the set of relevant ϕ-worlds with respect to w. So, the present framework assumes that the set of relevant antecedent-worlds is a function of the truth conditions of the antecedent. If two antecedents are necessarily equivalent, the corresponding sets of relevant antecedent-worlds will be the same.

[1] The formula $\ulcorner w \models \phi \urcorner$ is the formal expression of $\ulcorner \phi$ is true at $w \urcorner$.

The selection function f should be subjected to a variety of constraints. An obvious one is this:

Antecedent-Worlds. Let w be any world and A a set of worlds. Then

$$f_w(A) \subseteq A.$$

Simply put: Relevant antecedent-worlds are antecedent-worlds. If $A = \emptyset$, i.e. the antecedent is impossible, the constraint implies that $f_w(A) = \emptyset$, i.e. that there are no relevant antecedent-worlds. This seems plausible.

If, however, there are worlds satisfying the antecedent, we would like to secure the existence of relevant antecedent-worlds. A straightforward way to achieve this is to require that the selection function selects some antecedent-worlds unless there the antecedent is impossible:

Non-Emptiness. Let w be any world and A a non-empty set of worlds. Then

$$f_w(A) \neq \emptyset.$$

Another plausible constraint is known under the label 'weak centering' (cf. Lewis 1973: 29).

Weak Centering. Let w be any world in A. Then

$$w \in f_w(A).$$

In words: If the antecedent is true at the world w of evaluation, then w is a relevant antecedent-world. This constraint flows naturally from the idea that relevant antecedent-worlds are worlds which come from the world of evaluation by integrating the antecedent at small costs. If the antecedent is already true at the world of evaluation, the antecedent can be integrated at no costs at all.

The following constraint is sometimes added as well (we will further discuss it in section 6.1.3):

Equivalence for Relevance. Let w be any world, A and B be sets of worlds. If $f_w(A) \subseteq B$ and $f_w(B) \subseteq A$, then

$$f_w(A) = f_w(B).$$

In words again: If the relevant A-worlds are B-worlds and the relevant B-worlds are A-worlds, then the relevant A-worlds coincide with the relevant B-worlds. Thinking of the relevant worlds as in some sense being the closest worlds would motivate this constraint, for if the closest A-worlds are B-worlds and the closest B-worlds are A-worlds, then these sets of worlds must be the same.

On the present conception, the set of relevant A-worlds will usually contain many worlds. In contrast, Stalnaker (1968) assumed that there is always exactly one relevant A-world, with a possible exception for the case where A is impossible, i.e. the empty set. More precisely, we are looking at this *uniqueness assumption*:

Uniqueness. Let w be any world in $\mathcal{W}$ and $A \subseteq \mathcal{W}$. If $A \neq \emptyset$, then $|f_w(A)| = 1$.

Uniqueness tells us that there is always exactly one relevant A-world. If one assumes that relevance goes with what is most similar or closest, the assumption is that there is always a single most similar or closest world. Consequently, there cannot be two equally close worlds or a never-ending chain of closer and closer worlds. We will have to say more about the uniqueness assumption below.

A straightforward consequence of the uniqueness assumption and weak centering is what has been called 'strong centering':

Strong Centering. Let w be any world in A. Then

$$f_w(A) = \{w\}.$$

Weak centering tells us that if A is true at the world of evaluation w, then w is a relevant world. But if there cannot be more than one relevant world, then the only relevant world must be w. Strong centering is not implausible even without the uniqueness assumption. The constraint may be derived from the thought that a world is closer to itself than to any other world and so must be the closest antecedent-world in cases in which the antecedent is already true at it. Note also that weak centering is an immediate consequence of strong centering.

3.2 Relevance and Chance

Let us consider a probability function Ch_w over the whole set of possible worlds $\mathcal{W}$.[2] Given sets of worlds A and B, we may call the conditional

[2] As far as the underlying algebra of propositions is concerned, I assume that it is the powerset of $\mathcal{W}$. Ultimately, this may not be a realistic assumption. We should at least reckon

probability $\mathrm{Ch}_w(B|f_w(A))$ the *counterfactual chance of B under A* (relative to w). Here we conditionalize a chance function not on the proposition A, but rather on the set of relevant A-worlds $f_w(A)$ (which will in general be a proper subset of A). For worlds w and propositions A, we may write $\mathrm{Ch}_w^A(\bullet)$ for $\mathrm{Ch}_w(\bullet|f_w(A))$ to have an expression for the counterfactual chance given a possible antecedent A.[3] We presuppose that A is not the empty set, for otherwise the conditional chances would be undefined (more on definedness below).

The counterfactual chance function weighs the relevant A-worlds. It describes how likely B is made by the relevant A-worlds. Thus, it adds a more fine-grained structure to the set of relevant A-worlds. Where the selection function provides only information about which worlds are relevant, the chance function adds information about how relevant each world is: some worlds may count more than others. Although we have defined counterfactual chances in terms of an absolute probability function Ch_w, only the conditional probabilities given sets of relevant worlds will matter, so one could also have started with these conditional probabilities directly (we come back to this below). It is of course somewhat arbitrary to call this notion 'counterfactual chance'. The label would also have fitted a notion which is directed at the (physical) chance the consequent would have had if the antecedent had been true. But this is not what we shall subsume under the label 'counterfactual chance'.

Concerning the nature of this kind of chance, we may construe the chance function as concerning objective matters of fact just like we may take relevance to concern objective relations between possible worlds. But if one prefers, one could probably also see the chance function as encoding certain projections stemming from our practice of counterfactual reasoning in contexts of uncertainty (cf. the picture in Stalnaker 1984).

with the possibility of chance gaps, that is propositions for which the chance function is not defined. It could be that some sets are non-measurable for formal reasons (note, though, that we may only have to presuppose finite additivity for Ch_w) or it could be that for certain propositions no plausible chance-value can be found.

[3] The functions $\mathrm{Ch}_w^A(B)$ corresponds to the functions Gärdenfors (1988: 112), Joyce (1999: 197f.), and Lewis (1981) use to generalize the notion of imaging (more on this below). Others have considered similar functions, too. Following Bigelow (1976), who derives probabilities from a similarity metric over possible worlds, Bennett (2003: 248f.) introduces a notion which he calls *relative probability*. Sobel had a conception of counterfactual chances as well (cf. the report in Lewis 1981: 16–18).

Such a notion of counterfactual chance has the right profile to accommodate the constraints which emerged from our discussion of Skyrms's proposal. By having as their domain all metaphysically possible worlds, counterfactual chances are well defined for antecedents which contradict the actual laws of nature or the actual physical probabilities. As counterfactual chances are not identified with physical chances, they can be assumed to be non-trivial even when evaluated from the perspective of a deterministic world. And by being conditionalized on the set of relevant antecedent-worlds rather than, say, the set of all antecedent-worlds, we are free to assume that with respect to Morgenbesser cases only those worlds are relevant at which certain cotenable facts obtain.

The counterfactual chance function is presently relativized to possible worlds. However, given that the selection function is already relativized in this way, it is not clear whether this is really necessary. For different worlds, the selection function may already take us to different sets of relevant worlds and will thus make sure that the counterfactual chances can be different. Relativizing the chance function to possible worlds would only be necessary if there could be two possible worlds w and w' such that $f_w(A) = f_{w'}(A)$, but for some B, the counterfactual chance of B under A would be different for w and w'. I cannot think of an example which would indicate that we might need this extra complexity, but neither do I know of any good reason why there should not be any. So, it is merely in order to simplify our discussion that I am mostly going to ignore this possible relativization.

Let us see what more can be said about counterfactual chances. A basic feature they should have is that the counterfactual chance of a proposition under itself is always 1. This is already a consequence of the corresponding constraint for the selection function. Given that relevant A-worlds are always A-worlds, any probability function conditionalized on $f_w(A)$ will assign probability 1 to A, for $f_w(A)$ is a subset of A. So we find:

Chance 1. For all $w \in W$ and all non-empty $A \subseteq W$:

$$\mathrm{Ch}_w(A|f_w(A)) = 1.$$

The above constraint tells us somewhat trivially that A should be accepted if it is entertained counterfactually, i.e. $\mathrm{Ch}_w^A(A) = 1$. We may also note that the weak centering condition will have the consequence that if A is true at w, the world w can be given some weight under the counterfactual supposition of A. If strong centering is assumed, then the counterfactual chance becomes extreme for cases in which $w \models A$: it will assign 1 of 0 to any proposition B depending on whether $w \models B$, for Ch_w^A would come from conditionalizing on the singleton set $\{w\}$.

So far, we have looked at some immediate consequences for counterfactual chances coming from constraints on the selection function. Let us now turn to some constraints one may wish to impose on the chance function directly. As counterfactual chances are defined by conditionalizing on the set of relevant worlds, only relevant worlds can be assigned positive counterfactual chance. A non-relevant A-world will never be given any weight under the counterfactual supposition of A. Now, the converse claim seems desirable as well: every relevant A-world should be given some weight. So, if w is a relevant A-world, then w should be assigned positive chance under the counterfactual assumption of A. In a way, this may seem to be the whole point of the conception of relevance. A relevant A-world is a world which we should, on ideal reflection, take into consideration in evaluating what would be the case if A were true. Therefore, it seems that the counterfactual chances should not ignore any of the relevant A-worlds.

More formally, we are are looking at the requirement that

(*) $\mathrm{Ch}_w(\{w'\}|f_w(A)) > 0$, for all non-empty sets A and all $w' \in f_w(A)$.

Any relevant A-world (strictly speaking the corresponding singleton set) would receive positive chance under the counterfactual supposition of A.

A slightly stronger condition, known as *regularity*, would be that $\mathrm{Ch}_w(\{w'\}) > 0$ for all $w' \in \mathcal{W}$. That is to say, our chance function defined over all possible worlds would not ignore any world. If this condition holds, the chance function will assign positive weight to $\{w'\}$ under any set of possibilities A containing w', for this can only raise but not lower the probability of w'. The converse does not hold: condition (*) does not imply regularity, for the selection function may happen to avoid selection of counterexamples to regularity. It would imply regularity if (*) is required to hold for all possible selection functions, or a sufficiently rich set thereof. As far as I can see, the range of possible selection functions we are presently envisaging, that is those which would provide a realistic model of the notion of relevance, is rich enough to inherit any possible problems with regularity (more on this below).

Regular probability functions assign positive probability to any possibility, that is to any element $\{w\}$ in their domain. As a consequence, any non-empty set is assigned a positive probability. So, for regular functions, possibility and positive probability coincide. We could then even drop the notion of a relevant A-world and simply define a relevant A-world as a world receiving positive counterfactual chance under A. This would mean that the present theory would only have one primitive notion, the concept of counterfactual chance, with relevance defined in terms of it, where there are currently two independent, though interrelated, notions.

The problem with regularity is that it cannot always be satisfied by standard probabilities in infinite settings. The issue is typically illustrated with the so-called *dart board problem*. Suppose there is a dart board with infinitely many points on it and that a dart is thrown at the board in a completely random fashion. We would like to say (a) that it has a positive chance of hitting any of these points and (b) that for any point the chance is the same. But this is impossible in standard probability theory. We cannot assign the same positive probability to all the disjoint events described by 'the dart hits point *i*', for their disjunction would then sum up to a probability greater than 1 given that there are infinitely many points *i* (for more discussion, see Williamson 2007).

Regularity does not cause any problems in finite settings. In infinite settings, it may be very hard if not impossible to articulate in any plausible way a sense in which counterfactual chances should not ignore relevant worlds.[4] Now, if one allows the set of relevant worlds to contain more than one world, typically many worlds, I see no reason why this set could be assumed to be always finite (actually, it might be more realistic to suppose that it is in most cases infinite). To make the point, one could consider an infinite lottery which did not actually take place. What if it had? It seems plausible that there is for any of the infinitely many tickets a relevant world in which that ticket wins. So there would be infinitely many relevant antecedent-worlds. If one were to require, which may seem initially plausible, that all these relevant worlds should receive the same positive chance, then one would look at the same structure as in the dart board problem. For this reason, I do not think that (*) can be upheld in standard frameworks for probability.

[4] See Haverkamp and Schulz (2011) for how a regularity constraint might be implemented in a comparative setting.

As a matter of fact, I do not think that one has to invoke infinite lotteries in order to make the point. Just think of a standard coin toss case. If the coin had been tossed, which curve in space would its course have described? Arguably, there are infinitely many possible and equally relevant candidates, differing only in minuscule detail.

In the light of the fact that regularity is problematic in infinite settings but unproblematic in finite environments, it is possible to weaken the constraint (*) by restricting it to finite cases:

Limited Conditional Regularity. If A is non-empty and $f_w(A)$ finite, then $\mathrm{Ch}_w(\{w'\}|f_w(A)) > 0$ for all $w' \in f_w(A)$.

The limited constraint applies when the set of relevant antecedent-worlds is finite. In such situations, every relevant world is assumed to receive positive counterfactual chance. This avoids the problems with regularity in infinite settings, but if the suspicision is correct that sets of relevant antecedent-worlds will rarely be finite, the weakened constraint helps shaping our picture of the counterfactual chance function only to a small extent.

There are two further issues concerning the definedness of the chance function in certain situations. An immediate question is: What if the counterfactual assumption is impossible? More precisely: what should be the value of $\mathrm{Ch}_w^A(B)$ if A is the empty set? We know from the previous section that the set of relevant A-worlds, $f_w(A)$, is empty if A is empty. This is as it should be: if there are no A-worlds, there can be no relevant A-worlds. Given that we have defined counterfactual chances by conditionalizing on the set of relevant worlds, we would conditionalize on the empty set which always has probability zero. But according to the standard definition of conditional probabilities (see section 2.1), they are undefined for zero-probability sets. Hence, counterfactual chances will be undefined for impossible assumptions. I suggest setting the counterfactual chances to 1 in this case. So, we would make the following stipulation about counterfactual chances:

Vacuity. For all $w \in W$ and all $B \subseteq W$: $\mathrm{Ch}_w^{\emptyset}(B) := 1$.

This constraint parallels the idea that counterfactuals are vacuously true if the antecedent is impossible.

For all that has been said so far, the impossibility of an assumption may not be the only reason why counterfactual chances could go undefined. It might also happen that the absolute probability $\mathrm{Ch}_w(f_w(A))$ of a non-empty set of relevant worlds is zero. In this case, the counterfactual chance conditional on this set would, according to the standard definition, be undefined. This might be an unwelcome result if this set of worlds corresponds to a counterfactual supposition we are capable of entertaining. So, if there is a non-empty set of relevant A-worlds, the counterfactual chance conditional on A should, if possible, be defined. One condition which would secure this is the regularity constraint mentioned above, for it requires that only the empty set is assigned probability zero. As regularity is unproblematic in finite settings, the present issue is only a possible problem in infinite domains. Perhaps the best way to deal with it in this case would be to take conditional probabilities to be primitive and represent them with something like Popper measures (cf. Hájek 2003). We could then assume that the conditional probabilities given sets of relevant antecedent-worlds are mostly defined even for zero-probability sets. Moreover, this would square well with the fact that our grip on counterfactual chances comes from our judgements about the conditional probabilities $\mathrm{Ch}_w(B|f_w(A))$. We do not seem to have any direct access to absolute probabilities of the form $\mathrm{Ch}_w(B)$.

Lastly, let us consider a possible constraint which would simplify the notion of counterfactual chance greatly. It may seem that in evaluating a counterfactual supposition, we simply estimate the fraction of consequent-worlds among the relevant antecedent-worlds (Bennett 2003: 248f. seems to make this assumption). For example, it may seem that we think it about 1/2-likely that the coin would have landed heads under the counterfactual supposition of it having been tossed because we take about half of the toss-worlds to be worlds at which the coin lands heads. This line of thought would lead to the following uniformity condition:

Uniformity. Let $w \in W$ and $A \subseteq W$. If $f_w(A)$ is finite and non-empty, then for all $w \in W$ and all $B \subseteq W$: $\mathrm{Ch}_w^A(B) = \frac{|B \cap f_w(A)|}{|f_w(A)|}$.

The set of relevant A-worlds, that is $f_w(A)$, needs to be finite and non-empty for the fraction to be well defined. As with the regularity constraint, there are technical problems with finding plausible extensions of the uniformity condition which cover the infinite case.[5] I will not go into details about possible options.

[5] One problem is that if a probability function over a countably infinite set is countably additive (the probability of a countable union of disjoint sets is the infinite sum of the single probabilities), then there is no uniform distribution over this set.

According to uniformity, the counterfactual chance of B under A would be the fraction of B-worlds among the relevant A-worlds. Basically, uniformity requires that every relevant antecedent-world (or every set of worlds of the same size) is assigned the same weight. We will further discuss and tentatively reject the uniformity constraint in section 6.2. It will become apparent that thinking about proportions of consequent-worlds is probably only a good approximation to how counterfactuals are evaluated, but ultimately the real work needs to be done by a conception of counterfactual chances which is not reducible in such a simple way.

3.3 A Principal Principle for Counterfactuals

It seems that our credences in counterfactuals aim at the corresponding counterfactual chances. They specify a kind of ideal for what to think about counterfactuals. On the assumption that the counterfactual chance of the consequent under the antecedent is so-and-so, our subjective probability of the counterfactual should be so-and-so. As it turns out, this yields an evaluation constraint on counterfactuals similar to Lewis's Principal Principle which says that conditional on the assumption that the objective chances of a proposition are so-and-so, our credence in the proposition should be so-and-so. A Lewis type formulation of the evaluation constraint is merely a different way of representing the subjective probabilities of counterfactuals as weighted sums of chance hypotheses as Skyrms did (modulo, as we shall see, admissible evidence). Hence, despite the difference in form, we continue to look for a sense in which the subjective probabilities of counterfactuals can be explicated as an estimate of a more objective kind of chance, with chance now being interpreted as counterfactual chance.

I will simply start by presenting the evaluation constraint in the form of the Principal Principle and then explain it step by step. Recall that $[\![\bullet]\!]$ is an interpretation function which assigns to any sentence ϕ of a given background language the set of worlds $[\![\phi]\!]$ at which the sentence ϕ is true. Recall further that a conditional is simple if it does not embed another conditional.

PP Constraint. Let $\ulcorner\phi > \psi\urcorner$ be a simple counterfactual, let $w_@$ be the actual world, let x denote a number in the unit interval, and let Cr

be a rational initial credence function. Moreover, let X be the sentence $\ulcorner Ch_{w_@}^{[\![\phi]\!]}([\![\psi]\!]) = x \urcorner$ and let E be admissible. Then

$$Cr(\ulcorner \phi > \psi \urcorner | \ulcorner E \wedge X \urcorner) = x,$$

whenever the conditional probabilities are defined.

To simplify, let us ignore the role of admissible evidence for a moment; we will come back to it in the next section. The evaluation constraint is concerned with the conditional subjective probabilities of an initial epistemic state. Recall that an initial credence function represents an epistemic state which has not yet acquired any empirical information. For such an epistemic state, the evaluation constraint requires the following: conditional on the assumption that the corresponding counterfactual chances are x, we should take the counterfactual to be x-likely.

As we have seen, counterfactual chances are relativized to worlds. Hence, in the evaluation constraint, a world needs to be specified relative to which the counterfactual chances are taken. The obvious choice is the actual world: when thinking about the counterfactual chances, we are interested in the counterfactual chances with respect to the actual world. This is an instance of the more general fact that when we think about what is the case, we are interested in what is the case at the actual world. Besides, I have restricted the principle to simple counterfactuals, i.e. counterfactuals which do not embed another counterfactual, in order to set aside a discussion of nested counterfactuals for which a special story might be needed.[6]

In contrast to Lewis's Principal Principle, the present evaluation constraint for counterfactuals describes subjective probabilities as attaching to sentences. As mentioned earlier, this is a helpful way of circumventing the problem that counterfactuals may not have truth conditions. Taking credences to attach to sentences seems to be the most neutral way of expressing that counterfactuals are associated with certain subjective probabilities. To repeat, this is not meant to make an interesting claim about what the objects of credences fundamentally are.

A minor point. Sentences can be context-sensitive. To do justice to the possible context-sensitivity of sentences, one would need to relativize

[6] The question which would need addressing is whether the import/export principle as introduced by McGee (1985) is valid for counterfactuals. For some discussion, see Etlin (2008: 3.6).

credences to contexts when sentences are taken to be the objects of credences. So, one should speak of the credence a given epistemic subject assigns to a sentence *in a given context*. I have omitted this further complexity from the evaluation constraint.

3.4 Admissible Evidence

Let us now take a closer look at admissible evidence. Just as with the original Principal Principle, it is clear that some restrictions need to be imposed on which additional assumptions we can make over and above what the counterfactual chances are without altering our verdict about the counterfactual. On the assumption that a fully reliable oracle tells us that $\ulcorner\phi > \psi\urcorner$ is true, our credence in $\ulcorner\phi > \psi\urcorner$ may be raised to 1 and thus be different from what we might take the counterfactual chances to be. Even more clearly, the counterfactual under consideration will itself not be admissible. For instance, on the assumption that the coin would have landed heads if it had been tossed, our credence in this counterfactual should be 1 rather than 1/2, say, for the simple reason that $P(\ulcorner\phi > \psi\urcorner|\ulcorner(\phi > \psi) \wedge X\urcorner) = 1$ no matter what X is. Apart from $\ulcorner\phi > \psi\urcorner$, counterfactuals related to $\ulcorner\phi > \psi\urcorner$ may also be inadmissible, for they may imply the counterfactual, imply its negation, or else have an evidential import on it. In sum, not everything can be admissible.

It is worth observing that the notion of admissibility we need in the present evaluation constraint on counterfactuals is not the same as the one in the original Principal Principle. It may be helpful to introduce two separate labels at this point. Let *o-admissibility* be the notion of admissibility required for the original Principal Principle and let *c-admissibility* be the notion we will need for the evaluation constraint on counterfactuals. Perhaps the first thing to observe is that *c*-admissibility is not time-relative. *O*-admissibility is time-relative because objective chances are time-relative. If the objective chances are taken at a time t, then a piece of information E about a time t' after t will typically be inadmissible as long as it is not independent of the target proposition. However, it may become admissible when the chances are taken relative to a point t'' later than t'. In contrast, counterfactual chances are not time-relative: what the counterfactual chances are may vary from world to world, but it does not vary with time. Formally, this is reflected in the fact that the counterfactual chance function has a world-slot but not a time-slot. This is not to deny that counterfactuals may stand in an interesting relation to time. The antecedent of a counterfactual typically makes salient a point in time onwards from which the relevant counterfactual worlds are allowed to depart from the actual history. But this point is already fixed by the proposition expressed by the antecedent of the counterfactual, it it is not assumed to vary over time.[7]

A consequence of the temporal invariance of counterfactual chances is that whole world histories seem to be *c*-admissible, whereas only certain chunks of history are *o*-admissible at every time. Even if we know the whole history of a world but assume that only about half of the relevant antecedent-worlds make the consequent true, we should have an intermediate credence in the corresponding counterfactual. We need to be very cautious in making an assumption like this, though. On a semantic conception of possible worlds, on which a possible world closely corresponds to a totality of facts, the history of a world *w* should not be identified with the world *w* itself, for otherwise taking histories to be *c*-admissible would have the consequence that every *w*-truth is *c*-admissible, including truths about counterfactuals, because on a semantic picture of possible worlds, they decide all propositions. In particular, if a given counterfactual is true at *w*, it would itself be *c*-admissible. But as we saw above, this cannot be the case. In saying that whole histories are *c*-admissible, histories must be understood in a weaker sense according to which they only specify information about the course of spatio-temporal events taking place in the world.[8]

Information about the laws of nature seems to be both *o*-admissible and *c*-admissible. However, as far as the *c*-admissibility of objective chances is concerned, we have to be careful. The role of objective chance in the original Principal Principle is played by conditional counterfactual chance in the evaluation constraint for counterfactuals. And information about (conditional) counterfactual chance will be *c*-admissible just like

[7] One might want to allow the relevant point in time to be shiftable by the context of utterance. Even if this possibility were granted, this would not show that *o*-admissibility is time-relative in the sense intended here.

[8] It might be possible to think of histories as possible worlds in Lewis's (1986a) sense according to which they are maximally connected spatio-temporal objects. Such objects need not be assumed to decide all propositions, yet if one construes propositions as sets consisting exclusively of Lewis-worlds, they would do so.

information about (unconditional) objective chance is *o*-admissible. But the extent to which information about objective chance is *c*-admissible will be somewhat limited. To illustrate why, suppose that at a certain time *t*, it is determinate at a world *w* that a given counterfactual is true.[9] Then the objective chance at *t* of the counterfactual being true will be 1. This information cannot be *c*-admissible, for it would directly encode the truth-value of the counterfactual. Hence, particular information about objective chances of counterfactuals cannot always be *c*-admissible. Note that this does not exclude information about objective chances of basic non-counterfactual propositions. Neither does it exclude information about the conditional objective chances of the consequent of a counterfactual given its antecedent. What will sometimes be excluded is information about the absolute chance of a counterfactual being true.

As a matter of fact, it may well be the case that every piece of information which we can come to possess is *c*-admissible. For instance, it is doubtful that we can come to know that there is a fully reliable oracle which announces that the coin would have come up heads if it had been tossed (if there can be such an oracle at all). It may even be that there is no way for us to know whether the coin would have come up heads if it had been tossed (this will be discussed more thoroughly in chapter 6). So, although the assumption that the coin would have landed heads is inadmissible for the evaluation of various counterfactuals, in particular for the evaluation of 'If the coin had been tossed, it would have landed heads', it seems that we cannot come to know this assumption. However, even if the present conjecture that we cannot come to know *c*-inadmissible evidence is true, permitting only *c*-admissible evidence and not all kinds of information in the evaluation constraint will still be the strongest thing we can do, for the evaluation constraint is concerned with assumptions we may make. And we can suppose things to be true which we cannot come to know. But if *c*-inadmissible information is not in our epistemic reach, there will never be the need

[9] Recourse to determinacy is not necessary for making the point, but let me in any case be explicit about what I mean by it. I am assuming a conception of physical determinacy which is time-relative just like objective chance and largely behaves like chance 1 (determinate truth) and chance 0 (determinate falsehood). But due to problems with regularity, one cannot assume that chance 1/0 implies determinacy. For this reason, only the converse of this claim is a plausible constraint: if it is determinate at *t* whether *p*, then the objective chance of *p* at *t* is either 1 or 0.

to apply the notion of admissibility to a particular case which is concerned with our absolute, i.e. non-conditional, credences in counterfactuals. This would greatly diminish the importance of having such a notion. This issue will concern us again in sections 3.5 and 3.8 when we compare the present constraint with Skyrms's proposal and the imaging constraint, which are both formulated without reference to a notion of admissibility.

The fact that the original Principal Principle is stated in terms of the notion of admissibility is sometimes seen as a disadvantage, for it seems hard to get a sufficiently clear grip on this concept. (Recall, however, that utilizing the notion of admissibility is meant to generalize the principle. There is no problem in stating a weaker constraint without recourse to admissibility.) The same worry would apply to the present version of the evaluation constraint. (Note, however, that just as with the original principle, a weaker constraint without admissibility would be available by not permitting admissible evidence as a further conjunct.) Now, Lewis (1980: 96–8) already reformulates the Principal Principle in a way which frees it from the notion of admissibility and he further improves on this principle in Lewis (1994) to deal with the problem of undermining. I will briefly sketch this reformulation and then see whether we can make a similar move in the counterfactual case.

The basic idea is to explicate the relation between chance and credence exclusively in terms of histories and a theory of chance. We have already made use of histories: a history H_t describes a chunk of a world up to a certain time t such that H_t is true at two worlds w and w' iff they share their history up to t. A theory of chance T describes how the objective chances in a world depend on the history of the world up to any time t. So, for any history H_t, a theory of chance will tell us what the objective chances are at the end of H_t, i.e. at t. A theory of chance can be implemented in various ways. Lewis (1980) takes them to consist in the conjunction of all history-to-chance conditionals of the form $\ulcorner H_t \rightarrow P_t(A) = x \urcorner$, which tell us what the chances of a proposition A are at a time t when given a history up to time t as input and where '$\rightarrow$' is a 'strong conditional' of some sort, possibly a counterfactual as in Lewis (1973). Alternatively, one could think of T as specifying a global chance function for a world which delivers the chances at a time t when conditionalized on the history up to time t. If a particular history H_t is combined with a theory of chance, then this yields a verdict about the objective chances at t. Assuming further that histories and theories of chance are admissible, the original Principal Principle implies this simplification:

(1) $P_t(A) = Cr(A|H_t \wedge T)$,

where $P_t(A)$ is the chance of A at t as implied by $\ulcorner H_t \wedge T \urcorner$, H_t specifies a history up to t and T is a theory of chance. What this says is simply: conditional on a history and a theory of chance, we should adjust our credence in a proposition A to what the theory of chance says about the objective chances at t given H_t. Here we no longer use the notion of admissibility, although it is implicit in the justification of this variant that histories and theories of chance are admissible.

This variant of the Principal Principle has been turned into the 'New Principle' (see Hall 1994 and Lewis 1994) in response to the problem of undermining. The problem of undermining arises if we assume in a Humean spirit that chance supervenes on particular matters of fact. Then it might be possible that at a given time t a certain course of events F has a positive chance of occurring which, were it to occur, would undermine the chances at t, i.e. the chances at t would be different from what they actually are. Given that T would be undermined by F, T is actually incompatible with F, for F makes true a theory of chance incompatible with T. Hence, conditional on T, we should invest zero credence that F will occur. On the other hand, the Principal Principle suggests that we should have a positive credence in F, for F has a positive chance according to T. If undermining is possible, the latter principle seems to be false (see Thau 1994 for discussion of how the problem with undermining might generalize to non-supervenient conceptions of chance). The fix suggested by Hall (1994) and Lewis (1994) is to conditionalize the chances on the theory of chance, too. Thus we would get:

(2) $P_t(A|T) = Cr(A|H_t \wedge T)$.

The effect of this is to ignore undermining futures like F: conditional on T, they are no longer assigned a positive chance, for they are incompatible with T. Although there would be more to say here, this brief sketch should do, for we shall soon see that the possibility of undermining has no analogue in counterfactual chances.

Is there a principle similar in form to (1) which could substitute the evaluation constraint on counterfactuals? A natural way to approach this question is to see what the analogues of a history and a theory of chance might be in the case of counterfactuals.

As counterfactuals are not time-relative, there is no natural counterpart of a history in the evaluation of counterfactuals. Incidentally, the absence of time-relativity is also the reason why we do not have to worry about something like undermining. Without any futures, there are not going to be any undermining futures. Now, to some extent, the role of a history is played by the concept of relevance in the evaluation of counterfactuals (and which worlds are relevant depends to a large extent on the history of the world). The set of relevant worlds is what the counterfactual chances are relative to. If the relevant worlds share a common history, they can even be seen as specifying a history up to a time from which the counterfactual worlds are allowed to depart from the actual world. Let a theory of relevance be a theory which uniquely describes a selection function which provides for any world w and proposition A a set of relevant A-worlds relative to w.

It is evident that the role of objective chances in the original Principal Principle is played by counterfactual chances in the evaluation constraint on counterfactuals. In an analogous fashion, we may introduce theories of counterfactual chance which uniquely specify a chance function Ch which delivers, when conditionalized on a set of relevant A-worlds, the counterfactual chance of B under A. If we now take the two components together, we can reformulate the evaluation constraint on counterfactuals in a similar fashion.

So, let R be a theory of relevance and let T be a theory of counterfactual chance. Given propositions A and B, the conjunction of R and T will imply that the counterfactual chance of B under A is $\mathrm{Ch}^A(B)$, for some counterfactual chance function Ch (I omit a relativization to worlds). So, for any reasonable credence function Cr, we could require

(3) $\mathrm{Ch}^A(B) = Cr(A > B|R \wedge T)$.

Conditional on a conjunction of a theory of relevance and a theory of counterfactual chance, we should think about a counterfactual what the theory says about the counterfactual chance of the consequent conditional on the set of relevant antecedent-worlds. On the plausible assumption that a theory of relevance and a theory of counterfactual chance are admissible, this principle follows from the earlier PP-constraint just like the reformulation of the original Principal Principle follows from the latter. Thus, we gain a version of the constraint which no longer employs the notion of admissibility. In practice, considerations about admissibility are still going to be relevant, though, for in deciding whether the credence a given subject invests in a counterfactual can be justified by the principle, we do not get around the issue of deciding whether the subject possesses any inadmissible evidence.

3.5 Comparison with Skyrms

The PP-constraint closely mirrors the idea that evaluating a counterfactual amounts to an *estimate* of the corresponding chances. Let us assume that our present credences *Cr* come from an initial probability function *P* conditionalized on the evidence *E* we currently possess. Applied to a counterfactual, this gives us

(4) $Cr(A > B) = P(A > B|E)$.

Representing our present credences in terms of an initial credence function makes the PP-constraint applicable. We may now partition *E* according to the hypotheses H_x compatible with *E* which say for some *x* that the counterfactual chance $\mathrm{Ch}^A_{w@}(B)$ is *x*. Suppose this partition is finite. By the laws of probability, we then have

(5) $P(A > B|E) = P(A > B|\bigvee_x H_x) = \sum_x P(H_x|E) \cdot P(A > B|H_x)$.

In words: if we partition our evidence *E*, then in order to see how likely *E* makes a given statement, we may check how likely each partition cell makes this statement, weight this with the likelihood of the cell given *E* and take the sum over all cells.

Recall that H_x specifies that $\mathrm{Ch}^A_{w@}(B) = x$. Given that *Cr* comes from *P* by conditionalizing on *E*, we therefore have

(6) $Cr(\mathrm{Ch}^A_{w@}(B) = x) = P(H_x|E)$.

Suppose further that no H_x contains any inadmissible evidence. Then the PP-constraint is applicable to $P(A > B|H_x)$, which yields

(7) $P(A > B|H_x) = x$.

Taking all this together, we find

(8) $Cr(A > B) = \sum_x Cr(\mathrm{Ch}^A_{w@}(B) = x) \cdot x.$

So, in evaluating a counterfactual, we would estimate the counterfactual chances of the consequent given the antecedent by considering all possible candidate hypotheses about the counterfactual chances, see what they would prescribe, and weight the result by our credence that the hypotheses get it right. Although we needed a few steps to show that this follows from the PP-constraint, the fact that it follows is almost obvious. The PP-constraint tells us what to think under various hypotheses about the counterfactual chances. So, our unconditional judgement should be a weighted average of what the hypotheses prescribe.

Structurally, this is exactly Skyrms's constraint for counterfactuals with the notion of chance in his proposal being interpreted as counterfactual chance. So, modulo admissible evidence, the PP-constraint can be seen as a variant of Skyrms's proposal. We should note, though, that if it is possible to possess inadmissible evidence, then this variant of Skyrms's suggestion might give the wrong verdict, for it does not take such a possibility into account. It could then recommend uncertainty about a counterfactual despite the fact that we possess conclusive (inadmissible) evidence for or against the counterfactual. So, this variant can hold in full generality only under the proviso that it is not possible to acquire inadmissible evidence.

3.6 Imaging

There is an alternative route to an evaluation constraint on counterfactuals. It has a very natural motivation in term of suppositions (cf. Joyce 1999: ch. 6). Adams's thesis can be seen as describing the subjective probabilities of indicative conditionals as probabilities of the consequent arrived at under the *indicative supposition* of the antecedent. The idea would be that conditionalizing on an assumption adequately captures the effect an indicative supposition of the form 'Suppose such-and-such *is* the case' has on a belief state. Given such a thought, it is a natural question whether we can arrive at an interesting evaluation constraint on counterfactuals by investigating the impact a counterfactual supposition of the form 'Suppose such-and-such *were* the case' has on a belief state. It is clear that the two kinds of suppositions differ. We arrive at different belief states when we suppose indicatively that Oswald did not kill Kennedy or when we suppose counterfactually that he had not done it. Under an indicative supposition we hold on to what we already believe and merely add the indicative supposition to our stock of beliefs, whereas we often have to give up certain beliefs when integrating a counterfactual supposition.

Lewis (1976: 145–9) discusses an operation he calls *imaging* which provides an interesting model of the effect a counterfactual assumption might have on a belief state (he further elaborates on it in Lewis 1981). Lewis conducts his discussion in the context of a Stalnaker semantics for conditionals. For this reason, it is presupposed that the uniqueness assumption holds. This means that there is to any world w and proposition A (which is not impossible) a unique world w_A, which is the only relevant A-world. With the uniqueness assumption in place, it is not very hard to see how imaging works. The basic idea is simply that the subjective probability of w gets shifted to w_A. For an epistemically possible world w, we may ask what it would be like if A had been true. This will take us to w_A. For another world w', we may ask the same question. This will take us to w'_A. Doing this for all epistemically possible worlds, we arrive at a set of worlds w^i_A which are the epistemically possible candidates for how the world would have been like if A had been true. So, they are the worlds we should consider possible under the counterfactual supposition of A. How much weight should be given to one of these worlds? Exactly as much weight was given to the worlds it originated from as the single relevant antecedent-world.

To describe this formally, we follow Lewis in assuming, for simplicity, that the background set of possible worlds $\mathcal{W}$ is finite. We also set aside the case in which A is the empty set, i.e. the antecedent is impossible. Then given a probability distribution P over $\mathcal{W}$ representing a possible epistemic state, we can define the *image* P^A *of* P *under* A by setting

$$P^A(w') := \sum_{w:w_A=w'} P(w).$$

In this definition, the probability of a world w gets shifted to the world w_A. Given that a world can be the closest A-world for more than one world, a world w' may receive the probability of more than one epistemically possible world, which explains why we need to sum over all these worlds. Now, this defines the new probability of a single world w'. The probability of a whole set of worlds B is then given by $P^A(B) := \sum_{w' \in B} P^A(w')$, i.e. simply by the sum of probabilities over the worlds in B. The image of B under A, $P^A(B)$, then describes how likely B is under the counterfactual supposition of A. It is straightforward to verify that the image of P is again a probability function, for the probabilities are only redistributed among the worlds in $\mathcal{W}$.

Imaging seems to be a good candidate for describing the probabilities of counterfactuals. We would be looking at the hypothesis that

$$P(A > B) = P^A(B).$$

According to this idea, the subjective probability of a counterfactual is the probability of the consequent under the image of the antecedent. Structurally, this parallels Adams's thesis for indicative conditionals. The difference is that instead of the conditional probability of the consequent given the antecedent, we have the probability of the consequent under the image of the antecedent. We have already seen that conditionalizing and imaging describe different processes (see Lewis 1976: 148 for a numerical example).

The only problem with imaging is that it presupposes uniqueness. And the uniqueness assumption is, without any qualifications, not very plausible. Take a coin toss case with a fair coin and assume everything else is normal. Why should a heads-world be closer than a tails-world (or *vice versa*)? It seems that no matter how we understand the notion of relevance or closeness, it will allow for ties: two worlds whose differences regarding the actual world are perfectly symmetrical. To reiterate with an example by Lewis (1973: 80): is the closest world at which Verdi and Bizet are compatriots a world where Verdi is French or a world where Bizet is Italian? The two different ways of making them compatriots seem to require an equidistant departure from the actual world.[10]

It would therefore be nice to develop the notion of imaging without having to presuppose uniqueness. Simply dropping the uniqueness assumption would, however, put a hole in the definition of imaging: when we have a whole set $f_w(A)$ of relevant worlds, we can no longer shift the probability of w to a single closest world. What we would like to do, of course, is to shift it to the set $f_w(A)$. The question which arises is how the

[10] We are going to further discuss the uniqueness assumption in chapter 6. Perhaps I should warn the reader that although I am not going to adopt the uniqueness assumption for relevance, I will still defend a kind of Stalnaker semantics, so my doubts about uniqueness should not be taken to apply to Stalnaker type semantics in general. At this point, this mindset may seem somewhat paradoxical, but the air of paradoxicality will vanish once the general shape of the theory becomes clear.

probability shifted to the set $f_w(A)$ will be distributed over its members. Will any member be given equal share or are some worlds going to carry more weight than others? This is the same question which motivated the introduction of counterfactual chances. As we shall see in the next section, with the aid of counterfactual chances, imaging can easily be freed from the uniqueness assumption.

3.7 Generalized Imaging

Based on our conception of counterfactual chances, we can, as Gärdenfors (1988: 108–12) has shown, define a generalized form of imaging (see also Joyce 1999: ch. 6 and Lewis 1981: 15–18). As indicated above, the idea would be to shift the subjective probability of a world w to the set of relevant A-worlds $f_w(A)$ and then weight those worlds with the counterfactual chance $\text{Ch}_w^A(\bullet)$. Making a counterfactual supposition A would thus be described in the following way. For every epistemic possibility w, we would consider the set of A-worlds, relevant from the perspective of w, and weight these worlds with the counterfactual chance under A. In order to arrive at the ultimate suppositional belief state, we would perform this procedure for every epistemic possibility w and weight the results with the subjective probability of w. What to think under a counterfactual supposition A would depend on two factors: (i) what we take the actual world to be like and (ii) what we take the counterfactual chance under A to be for every way the actual world might be like. Although not simple, this way of describing counterfactual suppositions comes with a reasonable amount of phenomenological plausibility.

Formally, the *generalized image* P^A *of* P *under* A (I continue to use the symbol 'P^A' also for the generalized operation) can be defined by setting

$$P^A(w') := \sum_{w \in \mathcal{W}} P(w)\text{Ch}_w^A(w').$$

For simplicity, we assume here that the underlying set of worlds is finite, so that the sum is always defined (otherwise the constraint would have to be explicated in terms of integrals). As before, the probability of an arbitrary proposition B under the image of A is defined by simply summing over the elements in B, i.e. by setting $P^A(B) := \sum_{w' \in B} P^A(w')$.

Interpreted with generalized imaging in mind, the imaging constraint from the previous section would say that the subjective probability of a counterfactual is the probability of the consequent evaluated from the perspective of the generalized image of the antecedent. More precisely, the hypothesis would be this:

Imaging Constraint. Let $\ulcorner A > B \urcorner$ be a counterfactual whose antecedent is not impossible, i.e. $[\![A]\!] \neq \emptyset$, and let P be a subjective probability distribution. Then

$$P(A > B) = P^A(B).$$

Generalized imaging keeps all the virtues of imaging, but it no longer relies on the uniqueness assumption concerning relevant worlds. If we assume uniqueness, then the counterfactual chances will be extreme and assign probability 1 to the single most relevant A-world w_A and 0 to all other worlds, for they would come from conditionalizing on w_A. So, the chance term will reduce to 1 for $w' = w_A$ and is 0 otherwise, showing that generalized imaging does indeed generalize the original notion of imaging.[11]

Structurally, the epistemic state P^A representing what to think under the counterfactual supposition of A can be seen as the hypothetical belief state Adams and Edgington were after. Added to the picture are counterfactual chances.[12] Another difference is that the hypothetical belief state P^A makes the antecedent certain, i.e. $P^A(A) = 1$. This means that the idea of Adams and Edgington to evaluate a counterfactual like the corresponding indicative conditional in a related hypothetical belief state would get somewhat trivialized. With the antecedent A being certain, evaluating an indicative conditional $\ulcorner A \Rightarrow B \urcorner$ in P^A would reduce to merely evaluating the probability of the consequent B. This holds for the Adams/Edgington view on which an indicative conditional gets evaluated by the corresponding conditional probability, for if a probability function P is such that $P(A) = 1$, then $P(B|A) = P(B)$. But this reduction also holds for most other analyses of indicative conditionals. All that it pre-

[11] One thing to note is that in the absence of the uniqueness assumption, one has to impose the strong centering condition in order to keep probabilistic modus ponens. We will tentatively adopt the strong centering condition, but see 6.1.3 for how it might ultimately be relaxed.

[12] Edgington (1995: 321) comes close to suggesting that objective chances provide something like the *right thing to believe* about a given counterfactual. See also Edgington (2008: 15f.), where hypothetical belief states are compared to probability distributions over sets of relevant antecedent-worlds.

supposes is that the conditional satisfies modus ponens and is implied by the conjunction of antecedent and consequent, for then, given $P(A) = 1$, $P(A \Rightarrow B) \leq P(B)$ (by modus ponens) and $P(B) \leq P(A \Rightarrow B)$(by the conjunctive property). Thus, in no interesting sense would the evaluation of counterfactuals depend on the evaluation of indicative conditionals. What could be kept from Adams and Edgington is the idea that the evaluation of counterfactuals takes us to a hypothetical belief state by counterfactually supposing the antecedent.

3.8 Imaging and the PP-Constraint

There are two ways we can arrive at an evaluation constraint for counterfactuals. Imaging shifts the subjective probability of an epistemically possible world w to the set of relevant A-worlds relative to w, weighted by the counterfactual chance under A. Alternatively, we can require, in the form of a Principal Principle, that for every hypothesis that the counterfactual chances are a certain way, our credence in the counterfactual should be this way. The two approaches surely seem similar. Might they actually amount to the same thing? As we shall see, the answer is a qualified 'Yes': in the absence of inadmissible evidence and ignoring a special kind of uncertainty about counterfactual chances, the two constraints agree with each other. But without these assumptions, they may differ and when they do, it seems the Principal Principle type formulation is the safer bet.

Let us start with the role of admissibility. Here the situation is similar to how the PP-constraint compares to Skyrms's proposal. If it is possible to possess inadmissible evidence, the imaging constraint is in danger of making the wrong predictions. As mentioned earlier, however, it is unclear whether we can ever come to possess inadmissible evidence. Let me elaborate on this a little more. Recall that we are talking about c-admissibility here. It is clear that we can sometimes acquire evidence which is o-inadmissible at a certain time t. As time goes by, a lot of what used to be the future becomes known. It can therefore trump the information about the objective chances at t. But whether we can receive evidence about counterfactuals which can trump the information provided by the counterfactual chances is another question. An oracle might tell us that a counterfactual is true. Can this kind of testimony ever yield knowledge?

If knowledge by testimony requires knowledge by the testifying subject, the question would be how the oracle can know. Unless we postulate some magical access to counterfactual facts, even an oracle probably has to make a counterfactual supposition and see what follows. But then it would be in the same business as we are and can be expected to comply with the evaluation constraint.[13]

The issue of testimonial knowledge through oracles is not easy to decide, for the epistemic capacities of an oracle usually remain underspecified. Moreover, not every kind of oracle may be metaphysically possible. One could construe oracles as enhanced versions of human beings with perfect logical skills and radically more computational power. As a helpful analogy, we may apply this to the position of an epistemicist about vagueness holding that vague sentences express propositions which are true or false as the case may be but which are unknowable by epistemic subjects like us in situations like ours (cf. Williamson 1994). Although the facts registered by vague sentences are in a certain sense unknowable, the epistemicist is not committed to hold that these facts could not be known by oracles construed as enhanced versions of human beings, for the unknowability of those facts may owe itself to the various limitations of our epistemic capacities. The question would then be whether the case of counterfactuals is similar to the case of vagueness. Is it merely due to our limited epistemic capacities that we cannot know how the coin would have landed (assuming here that there is such a matter of fact)? If one answers 'Yes', perhaps because one thinks that the uncertainty about such counterfactuals actually stems from vagueness, then an oracle could probably tell us the truth-value of such counterfactuals. If one answers 'No', perhaps because one thinks, as I do, that there is a more principled barrier to knowing such counterfactuals (for more, see chapters 6 and 7), then even an oracle may not be able to know.

Testimony could also be a challenge for the imaging constraint in a less eccentric way. Suppose someone tells us that if a given coin had been tossed, it would have landed heads. Suppose further that she knows that the coin is some kind of trick coin which always comes up heads. We, on the other hand, do not know this. So antecedently to the testimonial

[13] If instead of knowledge we would talk about what can rationally be believed, the situation might change, for it might be that we can rationally believe an inadmissible proposition even if it is not an object of possible knowledge.

information we had—in accordance with the imaging constraint—a credence of about 1/2 in the counterfactual. Now suppose we have no reason to distrust our informant and therefore come to believe the counterfactual. As a result, our credence in the counterfactual will now be high or even be 1. In our new epistemic state, are we still satisfying the imaging constraint? One may think that we do not by assuming that the probability of the consequent under the image of the antecedent has not changed. Thus, one might conclude that we have acquired a piece of inadmissible information. However, the situation is more complicated than it may seem. The problem is that the information at hand also allows us to make indirect inferences about what the world must be like for the counterfactual being known by our informant. Based on our knowledge (or high credence) that our informant must have inferred the truth of the counterfactual somehow, we could conclude that the coin cannot be fair or else the chance process cannot have been a relevantly indeterministic one. But once we have updated also our more basic evidence concerning matters of fact which might ground the truth of the counterfactual, it might be that we are again satisfying the imaging constraint, for all ways of integrating the counterfactual supposition that the coin had been tossed will now be such that the coin is somehow determined to come up heads. It could thus be that the imaging constraint is in good shape even with respect to possible evidence from testimony. If not, we could always fall back on the more cautious PP-constraint.

Let us now turn to another possible difference. Imaging encapsulates the assumption that once we fix a world *w*, there can be no uncertainty about what the counterfactual chances relative to *w* are. For, in defining the image from the perspective of an epistemically possible world, we simply take the right counterfactual chances. But one may think that even given a world, we may still be uncertain about which worlds are the relevant antecedent-worlds and how they should be weighted. In contrast, the reformulation of the PP-constraint reveals that there does not seem to be a similar restriction on ways of being uncertain about the counterfactual chances, for the uncertainty about chances is a building block of this constraint. Whether this is a real difference might ultimately depend on our conception of a possible world in this context. The particular kind of uncertainty seems possible because we think of a world as something like a history of basic events or facts, which may leave it open in some cases what the counterfactual chances are. But possible worlds are often construed as entities which simply decide every question. Thought of in this way, they will also fix the counterfactual chances. Epistemologically, the first conception sometimes seems more natural, but the second one is the one standardly used in semantics and metaphysics. We do not have to decide this question. This is in any event at best a minor difference, which is not going to matter for what follows.

Modulo the two possible, but perhaps only apparent differences, the PP-constraint and the constraint in terms of imaging coincide. If we ignore admissible evidence, we only have to see that the imaging constraint is equivalent to the reformulation of the PP-constraint as an estimate of the counterfactual chances. We saw in section 3.5 that in this case the PP-constraint is equivalent to

(8) $Cr(A > B) = \sum_x Cr(\mathrm{Ch}^A_{w@}(B) = x) \cdot x.$

Compare this with the constraint from generalized imaging:

(9) $Cr(A > B) = Cr^A(B) = \sum_{w' \in B} Cr^A(w') = \sum_{w' \in B} \sum_{w \in \mathcal{W}} Cr(w)\mathrm{Ch}^A_w(w').$

Let us simplify for a moment the discussion by assuming that B is a maximally specific proposition corresponding to a single possible world. Then we can ignore the outer sum in the last term. Now, if we also set aside possible uncertainty about the counterfactual chances relative to a world, we can rewrite the credence in a chance hypothesis as the sum of the credences in the worlds making this chance hypothesis true:

(10) $\sum_x Cr(\mathrm{Ch}^A_{w@}(w') = x) \cdot x = \sum_x \sum_{w \models \mathrm{Ch}^A_w(w') = x} Cr(w) \cdot \mathrm{Ch}^A_w(w').$

As in every world exactly one chance hypothesis is true, the double sum in the second term sums over all worlds without counting any world twice. Hence, it simplifies to $\sum_w$, which shows that

(11) $\sum_x Cr(\mathrm{Ch}^A_{w@}(w') = x) \cdot x = \sum_w Cr(w) \cdot \mathrm{Ch}^A_w(w').$

This gives us the equivalence between the imaging constraint and the PP-constraint for the special case that the consequent of the counterfactual is maximally specific.

For the more general fact, we merely have to establish that the PP-constraint prescribes to sum over disjoint propositions in the consequent of a counterfactual, i.e. that the subjective credence in $\ulcorner A > (B \vee C) \urcorner$ for disjoint B and C is the sum of the credences in $\ulcorner A > B \urcorner$ and $\ulcorner A > C \urcorner$. Then summing in (11) over the worlds in B would indeed give us the credence of $\ulcorner A > B \urcorner$ as in (8). To see this, we merely have to recall that the counterfactual chance function is a probability function and is as such additive, i.e. sums over disjoint propositions. Now, take a theory of counterfactual chance, T. If we knew such a theory to be true, our credences in counterfactuals would, by the PP-constraint, match the values of the chance function and thus be additive too. But this also holds under the mere assumption of a theory T. Now, as our ultimate credences in counterfactuals can be represented—ignoring admissibility—as a weighted average of our credences in theories of counterfactual chance like T and the values of the resulting counterfactual chance function, this generalizes to the case where we are uncertain about the counterfactual chances. To put it another way: what holds under every assumption of a partition also holds unconditionally.

In conclusion, the two constraints are equivalent given the two provisos that admissible evidence can be ignored and that the particular kind of uncertainty about world-to-counterfactual-chance dependence does not occur. Given this equivalence, I will in what follows often speak indiscriminately about 'the evaluation constraint' meaning either of the two constraints.

3.9 Some Methodological Considerations

Before closing I would like to make a few methodological comments. The aim of the present chapter was to find a constraint which systematizes the intuitive data concerning our credences in counterfactuals. I think the imaging constraint and the PP-constraint, which are for all means and purposes equivalent, achieve this aim in a satisfactory way. They are explicated in terms of the two concepts of relevance and chance. Jointly, they define a notion of counterfactual chance which weights the relevant antecedent-worlds. Counterfactual chances can be seen as an enhanced notion of relevance which does not merely identify the relevant antecedent-worlds but also assigns them certain weights. According to this picture, relevance comes in degrees. As discussed earlier, one could have a conceptually more economical approach by starting with counterfactual chances and then define relevance as positive counterfactual chance. Because this strategy is as it stands limited to the finite case, I have opted for the slightly less economical approach which starts with two notions: a notion of relevance and a notion of chance. The idea in the background remains one of a graded conception of relevance.

Grading the notion of relevance might be a cost the present theory comes with, for it introduces a further concept, the notion of chance, in order to add additional structure to the notion of relevance. All things being equal, it would be better to be able to do without this extra complexity. But currently I do not see how this might be possible. The two different types of approaches, generalizing imaging and generalizing Skyrms's proposal in the form of a Principal Principle, both need a notion of counterfactual chance. The other existing accounts are either inadequate (Adams's prior probabilities) or involve a concept of similar complexity (Edgington's hypothetical belief states). So I think we have good inductive evidence that something like counterfactual chances are actually necessary to properly explicate the way counterfactuals are evaluated.

There is another worry one might have about the theory in its present shape. Although we laid down a number of formal constraints on the selection function and on the chance function, we have not said much about the actual content of these notions. Now, asking for a definition would be asking too much, for these notions will be part of the primitive building blocks of the ultimate theory. Every theory is based on some primitive terms. But let me try to say a little more.

The debate about the notion of similarity figuring in standard semantics for counterfactuals displays some general trends (see e.g. Lewis 1979, and Williams 2008 for some discussion in the context of objective chance). Although I prefer to use 'relevance' instead of 'similarity', the results matter just the same. So, for instance, we can say that relevant antecedent-worlds typically share most or even all of their history with the actual world up to a certain point shortly before the antecedent is made true. We can say that relevant antecedent-worlds do not violate the actual laws of nature from the time on at which the antecedent is made true (or even at no time at all if we follow Bennett's 1984 suggestion), provided that the antecedent is compatible with the laws of nature. We can also say that the antecedent is integrated at relatively small costs by preserving as many actual facts which are causally independent of it (the latter is something one would probably not want to say if one wishes to analyse causation in terms of counterfactuals). I do not go into detail about these issues, not because they would not matter to the present project, but because I feel I cannot add much to this debate. Whatever can plausibly be said about similarity as employed in standard semantics is something which can, by default, be said about the notion of relevance as employed in the evaluation constraints, too (there are some possible exceptions to this general rule which we shall have to discuss in chapter 7; see e.g. 7.4 for a discussion of the limit assumption).

Methodologically, the approach I favour for getting a grip on the desired notions of relevance and chance is to start with our credences in counterfactuals regarding particular cases and then ask what relevance and chance must be like to predict these credences via the evaluation constraints. This strategy is close in spirit to how Lewis conceived of how we should develop an adequate notion of similarity ('Analysis 2' in the quote to follow refers to his analysis of the truth-conditions of counterfactuals in terms of overall similarity of worlds):

> The thing to do is not to start by deciding, once and for all, what we think about similarity of worlds, so that we can afterwards use these decisions to test Analysis 2. Rather, we must use what we know about the truth and the falsity of counterfactuals to see if we can find some sort of similarity relation—not necessarily the first one that springs to mind—that combines with Analysis 2 to yield the proper truth conditions.[...] In looking for a combination that will stand up to the test, we must use what we know about counterfactuals to find out about the appropriate similarity relation—not the other way around. [Lewis 1979: 43][14]

One might worry that a theory set up in this way has only little predictive power. It is true that certain judgments about particular cases are not predicted by the theory, but rather needed to get a grip on the primitive notions the theory employs. Nevertheless, I think significant predictive potential remains. First, there is a feedback loop concerning particular cases. Although some judgements about counterfactuals are needed to inform the notion of counterfactual chance, the grip gained through this procedure can be used to make predictions about other yet unconsidered cases. Secondly, there is the possibility of explaining general facts about counterfactuals. Just like Lewis uses similarity to explain

[14] Page reference to the reprint in Lewis (1986b).

how counterfactuals can have the logic they actually have, our credences in counterfactuals become explainable once we avail ourselves of the notion of counterfactual chance. It is a non-trivial fact that the whole pattern of credences a subject assigns over a range of counterfactuals can be accounted for on the basis of her estimate of a single notion of counterfactual chance.

To see this strategy at work, let us consider the question of how counterfactual chances relate to objective physical chances. We have already seen in the course of our discussion of Skyrms's proposal that the two notions are unlikely to coincide. Counterlegals, for example, suggest that the needed notion of chance has a broader domain than physical chances. Morgenbesser cases might show that the chances cannot always be the conditional chances given the antecedent, but must include further information regarding cotenable facts. Finally, given that counterfactual chances are assumed to be non-trivial even in deterministic worlds at which there may not be any non-trivial basic physical chances, this is another reason why counterfactual chances should not be identified with (conditional) physical chances. Nevertheless, physical chances can still be expected to have an influence on counterfactual chances. And with respect to many simple cases, the two notions might even be in full agreement. For instance, the counterfactual chance that a coin would have landed heads under the assumption that it had been tossed may well equal the corresponding conditional physical chance.

There are hard cases which concern low-chance events which are, in a pre-theoretical sense, somewhat extravagant. Consider an example discussed by Hawthorne (2005):

(12) If the monkey had been put in front of a typewriter, it would not have produced *Hamlet*.

Assuming that all possible sequences of symbols the monkey might have typed in were equally likely, the problem is whether a world in which *Hamlet* is produced has to be counted as relevant. There is no general agreement on this matter: Hawthorne sees no way of privileging non-*Hamlet*-worlds over *Hamlet*-worlds, whereas Williams (2008) argues that *Hamlet*-worlds are not relevant because they are in a certain sense abnormal.

Do we have to side with Hawthorne on this matter? Although there is clearly a pull in this direction, this decision is not mandatory. The question turns on whether the counterfactual chances agree in such cases with the physical chances. Given that we do not have to identify the two notions, the present theory can accommodate either verdict. Following the suggested methodology, we would have to decide which credences are adequate with respect to the problematic counterfactuals.

If we count some *Hamlet*-worlds among the relevant antecedent-worlds, the theory predicts that we should take a somewhat qualified attitude towards such a counterfactual by only saying something like

(13) It is very likely that if the monkey had been put in front of a typewriter, it would not have produced *Hamlet*.

For in this case, the counterfactual chance would be high but not 1 and so the evaluation constraint would recommend a high credence somewhat shy of certainty. Taking this attitude would accord well with the intuitive thought that the monkey produces sequences in a completely random fashion, so it might just as well have produced *Hamlet*.

On the other hand, if we are inclined to rule out *Hamlet*-worlds, this will presumably be motivated by the impression that such a qualified attitude is unnecessarily modest. We could actually know that the above counterfactual is true. This would indicate that we should not consider *Hamlet*-worlds to be relevant. If *Hamlet*-worlds are excluded, the counterfactual chance given the the relevant antecedent-worlds will be 1.

Thus, according to the present strategy, the fine-tuning of the notion of counterfactual chance depends on finding the right epistemic attitude towards these kind of counterfactuals. I find it very hard to say whether such counterfactuals are knowably true or merely very highly likely, so I will not attempt to decide this question. What is important in the present context is only to see how the notion of counterfactual chance can be informed partly by our uncertain judgements about counterfactuals.

3.10 Conclusion

The aim of this chapter was to find a fully general constraint which describes the pattern underlying the ascriptions of subjective probability to counterfactuals. The key concept we introduced for understanding this relationship is *counterfactual chance*. It can be seen as the joint product of a conception of relevance and a conception of chance. Counterfactual chances can then be defined as the conditional chance of the set of consequent-worlds given the set of relevant antecedent-worlds. With counterfactual chances in place, there are two ways of explicating an evaluation constraint on counterfactuals. We can follow the structure of Lewis's Principal Principle and require the credence in a counterfactual to match the counterfactual chances conditional on any hypothesis about what the corresponding chances are. Essentially, we arrive at Skyrms's proposal with the notion of chance being interpreted as counterfactual chance. Alternatively, we can—following Gärdenfors (1988), Joyce (1999) and Lewis (1981)—generalize the notion of imaging to arrive at a notion of a *counterfactual supposition*. Imaging is a way of hypothetically revising our credal state according to a counterfactual supposition by shifting the credence for every epistemic possibility to the corresponding set of relevant antecedent-worlds weighted by the chance function. With minor provisos, the two type of constraints turn out to be equivalent. Whereas the Principal Principle type formulation is a variant of Skyrms's proposal, the evaluation constraint resulting from imaging is similar to the hypothetical belief state approach envisaged by Adams and Edgington, for the image under an assumption can be seen as a hypothetical belief state. Given the approximate equivalence between the two constraints we discussed in this chapter, it seems, then, that the two main trends in approaching the evaluation of counterfactuals converge to the same basic idea.

The formulation in terms of imaging is closest in form to Adams's thesis about indicative conditionals. Contrast

(14) $P(\ulcorner A \Rightarrow B \urcorner) = P(B|A)$

with

(15) $P(\ulcorner A > B \urcorner) = P^A(B)$.

We can see Adams's thesis as saying that the subjective probability associated with an indicative conditional is the probability of the consequent evaluated within the scope of an indicative supposition, where an indicative supposition is assumed to have the effect of conditionalization. In a similar vein, the imaging constraint requires the subjective probability invested in a counterfactual to be the probability of the consequent arrived at within the scope of a counterfactual supposition. The two constraints differ in the kind of supposition associated with the two kinds of conditional, but they share their general structure. Both equate the absolute probability of a conditional with the probability of the consequent relative to a supposition of the antecedent. Call this the *suppositional feature* of the two constraints.

In the case of indicative conditionals, the suppositional feature of Adams's thesis seems to be primarily responsible for why it is so hard to find a truth-conditional underpinning for this constraint, i.e. to explain the probability of an indicative conditional in terms of the probability of a proposition expressed by the conditional as a whole. Given that the evaluation constraint for counterfactuals shares the suppositional feature with Adams's thesis, differing only in having substituted a counterfactual supposition for an indicative supposition, we may expect similar problems in the counterfactual case.[15] Is the evaluation constraint compatible with counterfactuals having ordinary truth conditions or do we need an entirely different story in order to explain the intuitive credence assignments? It is fairly easy to see that the evaluation constraint cannot be explained in terms of the standard truth conditions for counterfactuals (cf. section 4.1 in the upcoming chapter). But what about other candidate truth conditions? These questions will concern us in the next chapter, where the challenge the evaluation constraint poses for an assignment of truth-conditions to counterfactuals will be analysed in more detail.

[15] Those familiar with Lewis's (1976) discussion of imaging will recall that the imaging constraint in terms of the non-generalized notion of imaging can be satisfied if counterfactuals are given Stalnaker's truth conditions. So, endorsing the uniqueness assumption for the selection function puts one immediately in a position to explain the imaging constraint. The problem becomes acute only when the notion of imaging is generalized in order to avoid the uniqueness assumption. We will come back to this point in our discussion of the triviality results in chapter 8.

4

A Puzzle About Counterfactuals

It is well known that indicative conditionals give rise to a puzzle concerning their truth conditions (see Edgington 1986, Edgington 1995: 278ff., and also Stalnaker 1975). Until very recently, the dominant assumption was that counterfactuals can stay out of such trouble. Modulo possible refinements and minor modifications, the standard truth conditions for counterfactuals seemed to be largely correct, while the door is wide open for how an adequate theory of indicative conditionals might finally look like. Some philosophers even take a radically bifurcated stance according to which indicative conditionals have no truth conditions whereas counterfactuals have something like the standard truth conditions (Bennett 2003, Gibbard 1981). Given that the two kinds of conditional are formed with the same connective, it seems, prior to investigation, somewhat unlikely that there is a big problem concerning truth conditions for indicative conditionals and nothing like that for counterfactuals.

The troublemaking data concerning the subjective probabilities of indicative conditionals are generated by the intuitive plausibility of Adams's thesis that indicative conditionals are evaluated by the corresponding conditional probabilities. Now, we have found in the previous two chapters that the intuitive way counterfactuals are evaluated is governed by a similar hypothesis which also brings into play conditional probabilities, albeit in a somewhat different way. It is a natural question whether the data captured by this hypothesis yield a similar problem for the assignment of truth conditions to counterfactuals. As we shall see shortly, this is indeed the case.

In the debate about indicative conditionals, there are two ways of identifying the problem of truth conditions. There is a simple and straightforward puzzle about the inferential properties of indicative conditionals concerning the question whether an indicative conditional $\ulcorner\neg A \Rightarrow B\urcorner$ follows from the corresponding disjunction $\ulcorner A \vee B\urcorner$. This is the puzzle discussed by Edgington and Stalnaker, which was mentioned above. In addition, there are the so-called triviality results initiated by Lewis (1976), which are formal results showing that Adams's thesis cannot be upheld on a truth conditional semantics once certain fairly weak assumptions are made. The puzzle and the triviality results are not unrelated. The puzzle can actually be seen as a less formal reformulation of a very basic triviality result. Given that the puzzle is much more accessible and it is far easier to determine what it shows and what it does not show, we will focus in this chapter on how a similar puzzle for counterfactuals can be generated. I will defer discussion of possible triviality results until later (see chapter 8), for they are better discussed against the backdrop of the theory I ultimately defend.

The chapter proceeds as follows. To motivate the problem, it starts with a discussion of how the evaluation of counterfactuals challenges standard semantics. Then I develop a more general puzzle about truth conditions for counterfactuals. Once the puzzle is identified, its background assumptions will be analysed and we will look at a possible means of escape. Finally, the puzzle for counterfactuals will be compared to the original puzzle for indicative conditionals.

4.1 The Problem for Standard Semantics

To get a feel for how the evaluation of counterfactuals challenges the possibility of assigning adequate truth conditions, let us consider the problem for standard semantics in some more detail. Recall that on the standard account, a counterfactual is true iff the consequent is true at all relevant antecedent-worlds. With the formalism developed in the previous chapter, we can state these truth conditions more precisely in the following form:

Standard Semantics. Let ϕ, ψ be sentences and $w \in W$. Then

$$(w \models \ulcorner\phi > \psi\urcorner) \equiv (\forall w' \in f_w([\![\phi]\!]) : w' \models \psi).$$

What this makes explicit is that the truth conditions are specified relative to a world w. The selection function then provides a set of relevant antecedent-worlds relative to w. With the universal quantifier governing the specification of truth conditions, we see that on the standard account, counterfactuals express a restricted kind of necessity, where the restriction is induced by the relevant antecedent-worlds (we come back to this aspect in the discussion of the duality thesis; see 7.2). Note also that on the standard account, counterfactuals with an impossible antecedent are always true, for in this case there are no relevant antecedent-worlds which makes the universal quantification vacuously true. With the standard truth conditions being explicated, we get a coarse-grained model of the proposition expressed by counterfactuals when we focus on the set $\{w : w \models A > B\}$, i.e. the set of possible worlds at which the counterfactual is true.

Here is an example Edgington (2008: 9) uses to illustrate the problem the standard account has with the evaluation of counterfactuals. A dog almost always but not quite always attacks and bites when strangers approach. Jacky does not approach. Based on the information we have, it seems we should take the following counterfactual to be likely but not certain:

(1) If Jacky had approached, she would have been bitten.

But this verdict of high probability without certainty cannot be predicted by the standard account:

> Arguably in all of my examples above [...], the counterfactuals would not come out as highly probable, but as known to be plain false on these truth conditions. Consider the dog that almost always bites when strangers approach. [...] It's not the case, and we take it not to be the case, that in *all* the relevant worlds in which I approached I was bitten. So the truth condition is not satisfied, and we believe it is not satisfied: we think it's certainly false that if you had approached you would have been bitten, according to the truth conditions. (Edgington 2008: 10)

Since the dog bites almost always but not quite always, there seem to be some relevant worlds at which he does not bite. As a consequence, the standard truth conditions are not satisfied. This shows that the counterfactual would be false according to the standard truth conditions. Moreover, the relevant epistemic subject should, according to the standard account, take the counterfactual to be false, since the subject can be assumed to possess all the empirical information needed to recognize that the standard truth conditions are not satisfied (in this case, this is most prominently the information that the dog not quite always bites when strangers approach). So, according to standard semantics, the subject should have a low credence, perhaps as low as 0, that the counterfactual is true. But this runs contrary to our inclination of assigning the counterfactual a high credence in the given epistemic situation.

A note of caution. The foregoing argument may appear to be a little quick. There seem to be ways of enriching the given story such that the standard account is in a position to explain the high credence we are inclined to assign to the counterfactual. Here is one such way. Suppose that whether the dog attacks depends in a complex but fully deterministic way on the clothing of the stranger. Most patterns of colour and shape make the dog attack, but there are some rare combinations which calm the dog down. Now, suppose that this kind of dependency is known. However, it is totally unknown which combinations of colour and shape make for aggression and which do not. In such a situation, the standard account seems capable of predicting a high credence in the counterfactual that the stranger would have been bitten had she approached. Why is this so? We should reason as follows. Either the stranger's clothes satisfy the pattern which cause the dog to attack or they do not. Since we know the general fact that most patterns cause aggression without knowing anything more specific, we should think it likely that the stranger's clothes are of the aggression causing type. If the clothing is of the aggression type, then we can assume that the dog attacks in all relevant worlds in which the stranger approaches, for whether the dog attacks depends only on the clothing. On the other hand, if the clothing is not of the aggression type, we can actually assume that the dog attacks in no relevant world in which the stranger approaches. Since the first case is most likely, we should think it likely that the dog bites in all relevant worlds in which the stranger attacks. Hence, we should think it likely without being certain that the standard truth conditions are satisfied.

This example indicates that there are cases in which the standard account is well equipped to predict the right amount of uncertainty about a given counterfactual. However, the specific way in which these cases are constructed prevents this line of argument from being generalized. They rely on the fact that the uncertainty derives from uncertainty about purely factual information and not from an only partial overlap of the relevant antecedent-worlds with the consequent-worlds (compare the preliminary discussion of possible sources of uncertainty in the introductory chapter). Hence, there are still plenty of ways of setting up the case which do not allow for a response along these lines (cf. Edgington 2008: 10). For example, think of the following two possibilities. To begin with, suppose that whether the dog bites is a completely random business: it simply attacks strangers with a high frequency, but selects its victims on a random basis. If this is so, then there are no unknown particular features of the actual world which either turn all relevant worlds into worlds at which the dog attacks or turn them into worlds at which it does not. Rather, worlds at which the dog attacks are equally relevant as worlds at which it does not. Therefore, only most but not all relevant worlds will be worlds at which the dog bites, and this will be knowably so. As a second case, suppose that whether the dog bites depends in a very subtle but deterministic way on the manner of approach. Although this time the selection of victims is a deterministic process, this does not necessarily help the standard account, since it may not be determinate in exactly which way the stranger would have approached if she had approached (see also the discussion in section 2.4.2). In many cases, the class of relevant worlds will contain various worlds in which the stranger approaches in slightly different ways reflecting her normal ways of moving. Most of these approaches will cause the dog to bite, but some of them pass by peacefully. Again, the standard account is forced to predict a low credence in the relevant counterfactual.

In sum, there are two ways in which the distribution of consequent-worlds among the relevant antecedent-worlds may be inhomogeneous. Either the counterfactual worlds can unfold in various different ways after the time the antecedent has been made true or there are already various different ways a world can make the antecedent true (see also section 1.1). If our uncertainty about a counterfactual is rooted in one of these two possibilities, the standard account has trouble explaining the uncertainty.

Having argued informally so far, it should be clear that a difference between the probabilities assigned to counterfactuals and the probabilities assigned to the proposition corresponding to the standard truth conditions is also a direct consequence of the evaluation constraint developed in the previous chapter. Suppose it is known that most but not all of the relevant antecedent-worlds are consequent-worlds. Then the counterfactual chances of the consequent given the antecedent are known to be high but less than 1. Plugging in the evaluation constraint, it follows that we ought to assign a high credence to the counterfactual. At the same time, we know that not all relevant antecedent-worlds are consequent-worlds, and so we are in a position to know that the standard truth conditions do not obtain. Hence, our credence that they obtain should be 0. If we stipulate '$\Box\!\!\rightarrow$' to have the standard truth conditions, i.e. as abbreviating the right-hand clause in their specification above, we see that the evaluation constraint implies that the credences in counterfactuals and in the standard truth conditions sometimes diverge:

(2) For some sentences A and B and a subjective probability function P: $P(\ulcorner A > B\urcorner) > P(\ulcorner A \Box\!\!\rightarrow B\urcorner)$.[1]

Now, we may ask whether, by the evaluation constraint, the probability of a counterfactual can sometimes be lower than the probability that the standard truth conditions obtain. To see that this is not the case, suppose that our credence in $\ulcorner A \Box\!\!\rightarrow B\urcorner$ has a certain value, x. Then, our credence that the counterfactual chance of A under B is 1 is at least x, for all relevant antecedent-worlds being consequent-worlds implies that the counterfactual chances are 1. If we now estimate our credence in the counterfactual via the evaluation constraint, we will assign probability 1 to it in a number of epistemic possibilities whose total probability is x. So our credence in the counterfactual will at least be x. This shows that the subjective probability of a counterfactual is bounded from below by the subjective probability that the standard truth conditions obtain:

(3) Always: $P(\ulcorner A > B\urcorner) \geq P(\ulcorner A \Box\!\!\rightarrow B\urcorner)$.

In a similar vein, we may ask whether we can also find a lower bound for the subjective probabilities of counterfactuals. As Joyce (1999: 199) observes, a lower bound is given by the subjective probability of $\ulcorner\neg(A \Box\!\!\rightarrow \neg B)\urcorner$. This is quite intuitive: if we invest some credence in a counterfactual, we should at least to the same degree assume that it is not the case that the negation of the consequent is a (restrictedly) necessary consequence of the antecedent by being true at all relevant antecedent-worlds. More formally, it is clear that if we invest some credence x in a counterfactual, this means, by the evaluation constraint, that our estimate of the counterfactual chances amounts to x. But on every chance hypothesis which assigns a positive value to B given A, the chance of $\neg B$ given A will not be 1 and so our credence in $\ulcorner A \Box\!\!\rightarrow \neg B\urcorner$ will be 0, from which

[1] I hope the fact that I use the symbol '>' ambiguously both as a sign for the natural language counterfactual and for the greater-than-relation does not cause any confusion.

it follows that our credence in $\ulcorner\neg(A \mathbin{\Box\!\!\rightarrow} \neg B)\urcorner$ will be 1. Hence, our total credence in $\ulcorner\neg(A \mathbin{\Box\!\!\rightarrow} \neg B)\urcorner$ will at least be $x \cdot 1$:

(4) Always: $P(\ulcorner\neg(A \mathbin{\Box\!\!\rightarrow} \neg B)\urcorner) \geq P(\ulcorner A > B\urcorner)$.

From the two inequalities, we may draw some preliminary conclusions about the inferential profile of counterfactuals. Given that the subjective probability of a counterfactual is always at least as high as the subjective probability that the standard truth conditions obtain but sometimes higher, we can infer that counterfactuals should be implied by but not imply the standard truth conditions. In other words, the truth conditions should be strictly weaker than the standard truth conditions.[2] Conversely, we can learn from the lower bound that the truth conditions of a counterfactual should at least require some of the relevant antecedent-worlds to be consequent-worlds, for otherwise they would not imply $\ulcorner\neg(A \mathbin{\Box\!\!\rightarrow} \neg B)\urcorner$.

A natural way of respecting these two lessons would be to assume that only a certain number of relevant antecedent-worlds have to be consequent-worlds. Bennett (2003: 249ff.) suggests such a weakening.[3] He calls it the 'near-miss proposal':

> **Near Miss.** A counterfactual is true just in case the fraction of consequent-worlds among the relevant antecedent-worlds is sufficiently high.

This way a counterfactual can be true even if not all relevant antecedent-worlds are consequent-worlds. At the same time, some antecedent-worlds must be consequent-worlds for the counterfactual to be true.

It is not hard to see, however, that theories like the near-miss proposal backfire (cf. Edgington 2008). Although they respect the two basic relations in which counterfactuals should stand to the standard truth conditions, they are not flexible enough to explain the whole range of credences in counterfactuals. Let us start by considering a case in which we do not take the number of consequent-worlds to be sufficiently high. We can

[2] Hájek (2007) provides arguments to the effect that counterfactuals actually have truth conditions stronger than the standard ones, suggesting that counterfactuals are strict conditionals. For how to accommodate at least one prominent argument of Hájek's, see section 7.2.

[3] An account somewhat similar in spirit but different in structure is developed by Williams (2008) in response to Hawthorne (2005).

assume that a fraction of about 50% is (by far) not sufficiently high. Now, we may think it about 50%-likely that if the coin had been tossed, it would have landed heads, because we take about 50% of the relevant antecedent-worlds to be consequent-worlds. However, this assignment of subjective probability cannot be justified on Bennett's proposal, for it should be obvious that the number of relevant consequent-worlds is not sufficiently high. On Bennett's proposal, we should rather be certain that the counterfactual is false. Secondly, consider a lottery which is so large that the fraction of losing tickets exceeds the threshold in Bennett's proposal. Then we can, according to Bennett's proposal, be certain that if we had bought a lottery ticket, we would have lost. Intuitively, though, we cannot be certain, for we cannot rule out that we would have won against all odds. Thus, it seems that Bennett's proposal over-generates certainty in some cases (where the fraction of relevant antecedent-worlds is above the threshold but below 1) but still under-generates uncertainty in other cases (where the fraction is below the threshold but above 0).[4]

Note that nothing seems to depend here on where we put the threshold in Bennett's proposal. Effectively, then, the proposal by Bennett covers almost all natural ways of weakening the standard truth conditions. This gets us already very close to the puzzle about counterfactuals, for it seems that the standard truth conditions are too strong while natural ways of weakening them do not work either. Let us now see how we can extract a precise argument out of this difficulty.

4.2 The Puzzle

We just saw that high subjective probability of a counterfactual is compatible with low subjective probability that the standard truth conditions obtain. Somewhat surprisingly, the situation changes in the case of certainty. Certainty about a counterfactual seems to require certainty that the standard truth conditions obtain. This is a consequence of the evaluation constraint. In order to be certain about a counterfactual, we must not give any credence to the possibility that the counterfactual chance of the consequent under the antecedent is less than 1, for otherwise our total credence in the counterfactual could not be 1, as it is a weighted average of

[4] For more problems with Bennett's proposal, see Edgington (2008: 12f.).

the values assigned by the chance hypotheses. So, we must be certain that the counterfactual chance is 1, which means that all relevant antecedent-worlds must be consequent-worlds. Hence, we must be certain that the standard truth conditions obtain. We therefore find:

(5) Always: If $P(\ulcorner A > B \urcorner) = 1$, then $P(\ulcorner A \Box\!\!\rightarrow B \urcorner) = 1$.

If we take the two features together, high probability of a counterfactual seems to be compatible with low probability that the standard truth conditions obtain, but certainty about a counterfactual seems incompatible with less-than-certainty that the standard truth conditions obtain. The first feature suggests that counterfactuals must be weaker than the standard truth conditions, but the second property seems to require that counterfactuals must be at least as strong as the standard truth conditions. In rough outline, this is the puzzle generated by the subjective probabilities of counterfactuals.

Our puzzle can be illustrated with the lottery example. In normal circumstances, we can be fairly confident about the counterfactual

(6) If I had bought a lottery ticket, I would have lost,

while at the same time even be certain that the standard truth conditions do not obtain. Let us now ask under which conditions we can know or be certain that this counterfactual is true. A typical situation would be one in which we know that the lottery was rigged in such a way that my ticket is left without any chance of winning. But now observe that in such a situation the standard truth conditions obtain and can be known to obtain. So, the picture changes for counterfactuals which are certain: we cannot be certain about a counterfactual without being in a position to be certain that the standard truth conditions obtain.

In addition, let us focus on a counterfactual which seems to be certain and show that it relates differently to the standard truth conditions in situations of uncertainty. Presently, we can be certain about the counterfactual

(7) If Obama had lost, McCain would have won.

Now, a typical situation in which we cannot be certain about this counterfactual would be one in which there is a promising third candidate. In such a case, we may only think it likely that McCain would have won if Obama had lost, because there are some relevant ways in which the third candidate could have won. But then we can actually be certain that the standard truth conditions do not obtain despite being justified in assigning a high credence to the counterfactual.

If we generalize these two observations, we arrive at the following puzzle:

The Puzzle. For most counterfactuals *c*:

(J) It is possible to possess a degree of justification less than 1 for *c* without being justified (not even indirectly) to the same or a higher degree that the corresponding standard truth conditions obtain.

(K) It is not possible to know *c* without being in a position to know that the corresponding standard truth conditions obtain.[5,6]

I present the puzzle here in somewhat more objective terms: knowledge and degrees of justification. It is equally possible to state the puzzle in terms of certainty and degrees of subjective probability. I prefer the more objective characterization because I take it to be more basic, but not much will hang on this choice. Let me also be explicit about the sense of possibility invoked in (J) and (K). It is simply meant to be metaphysical possibility. The same applies to the principles about epistemic access later on in this section.

I have restricted the puzzle to most counterfactuals, since certain trivial counterfactuals do not satisfy the first assumption. To see this, consider

(8) If someone had robbed the bank, someone would have robbed the bank.

It seems that we are always in a position to know that the standard truth conditions of this counterfactual obtain, for its trivial structure makes it plain that the relevant antecedent-worlds are always consequent-worlds. Moreover, we may always be fully justified in believing this counterfactual. But then there are already two reasons why (J) does not hold, for (J) requires the possibility of being uncertain about the counterfactual as well as the possibility of being uncertain about the standard truth conditions.

[5] I adopt a broad conception of being in a position to know comprising everything which can be known on the basis of our actual knowledge. Thanks to Miguel Hoeltje for this recommendation.

[6] For the sake of brevity, I often use phrases like 'to know *c*' in a slightly deviant sense (different from e.g. 'she knows John') meaning 'to know that *p*' for a '*p*' such that *c* = the proposition that *p*.

After having flagged this kind of exception, I will henceforth ignore, for ease of exposition, the need for such a restriction.

The puzzle about counterfactuals is a general challenge for assigning adequate truth conditions to counterfactuals. Assumption (J) suggests that the truth conditions for counterfactuals cannot imply the standard truth conditions. If they did, we could infer the standard truth conditions from the counterfactual and thereby transfer our justification for the counterfactual to the standard truth conditions. Thus, counterfactuals cannot have the standard truth conditions. On the other hand, (K) indicates that the truth conditions of counterfactuals should imply the standard truth conditions. Otherwise there would be a situation in which the counterfactual is true but the standard truth conditions do not obtain. Moreover, nothing seems to stand in the way of knowing that such a situation obtains. But then it would be possible to know that a counterfactual is true without being in a position to know that the standard truth conditions obtain for the simple reason that the standard truth conditions do not obtain.[7] So, no truth conditions weaker than the standard truth conditions seem to be adequate either. This, then, is the puzzle for counterfactuals: they seem to imply and not to imply the standard truth conditions.

The reasoning to the contradiction makes use of two principles linking epistemic attitudes towards propositions with their comparative strength.

Implication and Epistemic Access. Let *C* and *D* be propositions.

a. If *C* entails *D*, then it is not possible to be justified to a certain degree that *C* is true without being justified (at least indirectly) to the same or a higher degree that *D* is true.
b. If *C* does not entail *D*, then it is possible to know that *C* is true without being in a position to know that *D* is true.[8]

The rationale for these two principles is as follows. The first one is based on the idea that justification can be transferred by competent deduction. So, if *C* entails *D*, and we are justified to a certain degree that *C* is true,

[7] Here I assume factivity for being in a position to know: if we are in a position to know that *p*, then *p*.

[8] Edgington (1995: 279) discusses similar principles in application to indicative conditionals.

then competently inferring *D* from *C* can be supposed to transfer our justification for *C* onto *D*. The second principle may be rendered plausible by the following line of argument. If *C* does not entail *D*, then there is a possible situation in which *C* is true but *D* is false. Now, we may come to know that such a situation obtains. But then we would know that *C* is true without being in a position to know that *D* is true for the simple reason that *D* is false.

Before we further assess the plausibility of these principles, let me show how they provide the resources to give a direct argument to the effect that most counterfactuals lack truth conditions. Before that, a clarificatory remark about the terminology is in order. The question of whether conditionals have truth conditions is typically equated with the question of whether they express propositions. In this context, propositions are presupposed to encapsulate truth conditions so that we can assume a sentence to express a proposition just in case it has certain truth conditions. Moreover, both notions are construed in a *bivalent* way. With this in mind, an argument to the conclusion that most counterfactuals lack truth conditions is quickly at hand. Let *c* be a counterfactual which satisfies the two assumptions of the puzzle. To begin with, we find, by (J) and part (a) of the implication principle:

(9) If *c* has truth conditions, its truth conditions do not entail the standard truth conditions.

Suppose *c* has truth conditions. Then it expresses a proposition. By (J), it is possible to be justified to a certain degree that this proposition is true while not being justified to the same or a higher degree that the standard truth conditions obtain. But the first part of the implication principle directly implies that this could not be the case if *c* entailed the standard truth conditions.

Application of part (b) of the implication principle to (K) yields:

(10) If *c* has truth conditions, its truth conditions entail the standard truth conditions.

Again, suppose that *c* has truth conditions. Then it expresses a proposition. By (K), it is not possible to know that this proposition is true without being in a position to know that the standard truth conditions obtain. But then, by part (b) of the implication principle, the truth conditions of *c* must imply the standard truth conditions.[9] Finishing off the argument, the two conditional statements now jointly imply that *c* does not have truth conditions.

What can we do? In the following section, we will see that the background principles linking implication (or the lack thereof) with justification and knowledge probably do not hold in full generality. This provides a small leak in the argument one may be able to exploit. We will explore this possibility later in chapter 6.

4.3 A Means of Escape

How plausible are the general principles linking implication and epistemic access? I will sidestep discussion of some of their details which are important but do not strike me to offer a key for a possible resolution of the puzzle. To make the principles more precise, one would need to address (i) which conception of entailment is presupposed, (ii) which conception of propositions is assumed and (iii) how epistemic attitudes are taken to interact with propositions. For instance, on a conception of propositions as sets of possible worlds, entailment amounts to being a subset, but a very complicated story is needed about how epistemic attitudes attach to propositions so conceived to explain why we are often justified to have different epistemic attitudes in relation to sentences which are necessarily equivalent.

An analogue of the first principle is usually assumed in formal theories of justification and Bayesian epistemology. In standard probability theory, a logical consequence of a statement is never assigned a probability lower than the original statement. However, it should be noted that formal theories of justification are usually highly idealized. In particular, they tend to assign probability 1 to every logical truth. Real epistemic subjects are not logically omniscient and often need to undergo a cognitive process of a certain kind, of which competent deduction is a paradigm, to transfer their justification for a given statement to one of its logical consequences. I have tried to accommodate this fact by introducing the idea of indirect

[9] We may note that the assumption that the counterfactual *c* has truth conditions is only used to make the implication principles applicable to it. Opponents of truth conditions for counterfactuals would therefore need to explain why these principles lose their plausibility when applied to non-truth-conditional contents.

justification requiring only that logical consequences gain indirect justification from the justification of the statements from which they can be derived. To a good approximation, indirect justification may be construed as standing in a similar relation to justification as being in a position to know stands to knowledge. Still, the weakened principle probably needs further modification, but as far as I can see, this will not offer a promising route of escape to solve the puzzle.

For instance, one might approach the puzzle by holding that the truth conditions of counterfactuals are not transparent to ordinary speakers. If so, it could be that ordinary speakers evaluate counterfactuals in a way which diverges from how they would have evaluated them had they known about their truth conditions (which are, we may assume, the standard truth conditions). To some extent, such a line would constitute an error theory about counterfactuals. On the other hand, I agree that ordinary speakers are not semantically omniscient in the sense of being capable of specifying the truth conditions of counterfactuals in terms of relevant possible worlds, for example. For one, they may lack the conceptual resources to do so. For another, even upon aquiring the necessary concepts, developing a semantic clause remains a difficult task not everyone can successfully carry out. Fortunately, there seems to be room for steering a middle course between a full error theory and an assumption of semantic omniscience. Even if speakers lack knowledge of independently specifiable semantic clauses, they can still be expected, by default, to proportion their credences to the logical strength propositions have in relation to other propositions and to what they believe the world is like. Applied to counterfactuals, this would mean that if a speaker has knowledge of facts, say about a third candidate in an election campaign, which would clearly invalidate certain truth conditions, but the speaker nevertheless assigns the counterfactual a high credence, then this is, by default, evidence that counterfactuals do not have the proposed truth conditions.

The second principle strikes me as much more problematic. First of all, it is inconsistent with the existence of unknowable truths as expressed by Fitch-sentences. To see this, take C to be an unknowable truth. Since C is a truth, it is consistent and does not imply everything. Let D be a proposition not implied by C. Application of the second implication principle yields that it is possible to know that C is true without being in a position to know that D is true. However, the latter will not be possible for the simple reason that it is, by the choice of C, not even possible to know that C is true.

A similar point can be made on the basis of consistent but unknowable propositions, for the fact that Fitch-sentences are true was used above only to guarantee consistency.[10] As an example, consider 'There are no thinkers'. This sentence is false, but consistent, and so won't imply everything. By the second principle, it would be possible to know that there are no thinkers. However, this is impossible: if someone were to know that there are no thinkers, there would be a thinker, namely her, and so the proposition that there are no thinkers would be false, and hence could not constitute knowledge.

There is another source of counterexamples to this principle which does not involve unknowable truths. The principle is also challenged by knowable propositions which can only be known by being in a position to know a stronger proposition. Consider, for example, the following case. Let C be the proposition that snow is white and assume D to be the proposition that snow is white and there is a thinker. The proposition C does not imply the proposition D, for there are possible worlds in which snow is white but in which there are no epistemic subjects. But it is not possible to know C without being in a position to know D: if one knows that snow is white, one is always in a position to know that snow is white and there is a thinker, for the latter conjunct expresses a (contingent and perhaps a priori) truth which we are always in a position to know. So, this indicates that there are counterexamples to the principle which do not involve unknowable propositions but rather pairs of knowable propositions which differ in strength but not with respect to their knowability conditions.[11] It need not even be that D is strictly stronger than C. If D is any proposition which we are always in a position to know, such as 'There is a thinker' (or 'I exist'), then any C which does not imply D, such as 'Snow is white', will already constitute a counterexample.

There is also a formal argument by Rosenkranz (2004: 71) to the same conclusion. Consider a sentence with the following structure:

(11) $p \wedge (\mathcal{K}p \supset \mathcal{K}q)$,

[10] Thanks to Alan Hájek for pointing this out. The example which follows is also due to him.

[11] Cf. also Edgington (1995: 279) who reckons with certain counterexamples to this principle stemming from Moore-paradoxical sentences.

where the operator '$\mathcal{K}$' formalizes 'is known at some time by someone'. The crucial fact to observe is that

(12) $\mathcal{K}(p \wedge (\mathcal{K}p \supset \mathcal{K}q))$

implies $\ulcorner \mathcal{K}q \urcorner$ if we assume that the knowledge-operator distributes over conjunction, for in this case we find

(13) $\mathcal{K}p \wedge \mathcal{K}(\mathcal{K}p \supset \mathcal{K}q)$,

which, by the factivity of knowledge, further implies

(14) $\mathcal{K}p \wedge (\mathcal{K}p \supset \mathcal{K}q)$.

Now, this entails $\ulcorner \mathcal{K}q \urcorner$, while the original statement does not imply q. In sum, $\ulcorner p \wedge (\mathcal{K}p \supset \mathcal{K}q) \urcorner$ does not entail q, but knowledge of the former implies knowledge of the latter, i.e. it is not possible to know the former without knowing the latter. Therefore, the present case is also a counterexample to the second principle.

The failure of the second principle points to a possible means of escape from the puzzle. Suppose counterfactuals have truth conditions which are weaker than the standard truth conditions. But suppose further that our epistemic access to counterfactual truths is limited: we can only know that a counterfactual is true if we are in a position to know that a stronger proposition is true, namely that the standard truth conditions obtain. Such truth conditions could do justice to the first assumption of the puzzle, for counterfactuals would not imply the standard truth conditions. However, by having a more demanding epistemic profile, counterfactuals would satisfy the second assumption of the puzzle as well. Weaker truth conditions are not hard to find. The real question would be how counterfactuals could receive an epistemic profile with the desired properties. As remarked above, we will take up this question in chapter 6 after we have discussed in the next chapter whether the puzzle might actually show that the probability of a counterfactual should not be construed as a probability of the counterfactual taken as a whole.

4.4 Comparison with Indicative Conditionals

The puzzle about counterfactuals has been closely modelled after a corresponding puzzle for indicative conditionals. In this section, I will briefly make this connection explicit and comment on a few similarities and differences.

In the case of indicative conditionals, the puzzle concerns the so-called *'or'-to-'if' inference* (I got this label from Edgington 1995). The question is whether the following pattern is valid:

(15) A or B.

(16) Therefore, if not A, then B.

Basically, this inference states that indicative conditionals are implied by the corresponding material conditional, for exchanging $\ulcorner \neg A \urcorner$ for A and eliminating the double negation in the conditional, it suggests that $\ulcorner \neg A \vee B \urcorner$ implies the indicative conditional. Assuming further that indicative conditionals imply the material conditional (which they do as long as they obey modus ponens), indicative conditionals would have the material truth conditions. The plausibility of the 'or'-to-'if' inference thus supports the material analysis.

As we shall see, situations in which the material analysis struggles make the inference seem invalid. Stalnaker (1975) has concluded from this that the inference is only pragmatically valid but semantically invalid. Edgington (1986) (see also Edgington 1995: 278ff.) uses the Janus-faced behaviour of the inference to construct an argument to the effect that conditionals do not have truth conditions. Edgington's argument takes the form of a puzzle. It was the blueprint for the puzzle about counterfactuals developed in the previous section. Let us take a closer look at how the two puzzles relate to each other.

Despite its intuitive plausibility, the 'or'-to-'if' inference appears to be invalid when we focus on situations of uncertainty. Suppose there are three suspects: the butler, the gardener and the cook. It is very likely that the gardener did it and so quite unlikely that the butler or the cook did it. Still, the hypothesis that the cook did it is comparatively much more likely than the remote thought that the butler murdered his master. In this scenario, we should think it very likely that either the gardener or the butler did it (simply because it is already very likely that the gardener did it), but we cannot be certain as there is a chance that the cook committed the crime. Intuitively, we should not infer that if the gardener did not do it, the butler did, for if the gardener did not do it, the cook probably did it. This is strong evidence against the 'or'-to-'if' inference.

Interestingly, the inference looks much better when the disjunctive premise is certain, while both disjuncts remain at least somewhat uncertain.[12] To illustrate, suppose that the cook is not a suspect any longer. Then we can be certain that either the gardener or the butler did it. In this situation, we seem to be entitled to infer that if the gardener did not do it, the butler did. So, under conditions of certainty about the disjunction (but uncertainty about the disjuncts), the 'or'-to-'if' inference seems fine.

Just like the problematic features for counterfactuals derive from the evaluation constraint, the puzzle about the 'or'-to-'if' inference can be traced back to a feature of conditional probabilities: the probability of a disjunction $\ulcorner A \vee B \urcorner$ can be high while the corresponding conditional probability of B given $\ulcorner \neg A \urcorner$ is low, but if the probability of $\ulcorner A \vee B \urcorner$ is 1, the conditional probability of B given $\ulcorner \neg A \urcorner$ must be 1 as well. So if Adams's thesis is correct and indicative conditionals are evaluated by the corresponding conditional probabilities, then it could be explained why the 'or'-to-'if' inference fares well in situations of certainty, but breaks down when the premise is uncertain even though it may deserve an arbitrarily high subjective probability.

In conclusion, Edgington's argument against truth conditions for indicative conditionals starts with two observations: (a) high subjective probability that the material truth conditions obtain is compatible with low subjective probability of the corresponding indicative conditional while (b) certainty that the material truth conditions obtain is incompatible with less-than-certainty about the corresponding indicative conditional.[13] The first assumption suggests that the material truth conditions are not as strong as the truth conditions of the indicative conditional. Otherwise we should always take the indicative conditional to be likely when we take the corresponding material truth conditions to obtain. In contrast, the second assumption indicates that the truth conditions of the indicative conditional cannot be stronger than the truth conditions of the material conditional. Otherwise there would be

[12] If one disjunct is certain as in '$2 + 2 = 4$ or $2 + 2 = 5$', the 'or'-to-'if' inference would force us to evaluate indicative conditionals with a certainly false antecedent, which is oftentimes a hard thing to do.

[13] If one is unsure whether indicative conditionals are true or even truth-evaluable when the antecedent is certainly false, then principle (b) should be qualified by requiring that the material conditional's certainty does not owe itself to the certain falsity of its antecedent.

a possible situation in which the material truth conditions obtain but the truth conditions of the indicative conditional do not obtain. Now suppose we become certain that such a situation obtains. Then we would be certain about the material truth conditions but uncertain about the indicative conditional contradicting feature (b). Conclusion: indicative conditionals show an epistemic behaviour which one would not expect if they had an ordinary truth conditional content. In a nutshell, this is how the 'or'-to-'if' inferences challenge truth conditional accounts of indicative conditionals.

The general structure of the problem for indicative conditionals is largely the same as the one we identified for counterfactuals. There are two differences, though. First, indicative conditionals are compared with the material conditions, whereas counterfactuals are compared with the standard truth conditions for counterfactuals. Second, indicative conditionals are compared to truth conditions which are potentially weaker, whereas counterfactuals are compared to truth conditions which are potentially stronger.

Changing the truth conditions that counterfactuals are compared to is essential for developing a similar puzzle about counterfactuals. It is quite clear that the puzzle about indicative conditionals cannot be extended in a direct way to counterfactuals. For that to be possible, certainty about the material truth conditions would have to be sufficient for certainty about the corresponding counterfactual. But this is simply not the case. To see this, recall an example from chapter 2. Officer Shaker can be certain that either Linda or Elizabeth was at home last night, for he found a note saying 'I will be back tonight' which was written either by Linda or Elizabeth. However, Shaker cannot be sure that if Linda had not been home last night, Elizabeth would have been, for if Linda wrote the note, then Elizabeth probably had no inclination to come back early independently of Linda's whereabouts. For this reason, we cannot always be certain about a counterfactual on the basis of certainty about the corresponding material truth conditions.

Actually, much simpler counterexamples can be constructed from cases in which certainty about the material truth conditions is grounded either in certainty that the antecedent is false or in certainty that the consequent is true, for we can neither be certain about a counterfactual on the basis of being certain that its antecedent is false nor on the basis of being certain that its consequent is true. But since the corresponding hypothesis for indicative conditionals is sometimes restricted to cases in which certainty about the material truth conditions is not based on certainty about some of its constituents, it is important to note that there are counterexamples which could not be excluded by a similar restriction.

It might be that the fact that the puzzle for indicative conditionals does not directly affect counterfactuals has created the illusion that there is no problem of truth conditions for counterfactuals. For instance, Bennett (2003: 254ff.) takes this fact to show that there is no corresponding problem about truth conditions for counterfactuals. But drawing this conclusion has probably been a little premature. One may even wonder whether Bennett has missed the relevant dimension of comparison: in extending the argument to counterfactuals, one should not compare a counterfactual with the material conditional but rather with a proposition which relates to a counterfactual in a way similar to how the material conditional relates to the indicative conditional.[14] Counterfactuals do not stand in an interesting relation to the material conditional, but they do stand in an interesting relation to the standard truth conditions.

4.5 Conclusion

In this chapter, we have taken a closer look at how the data about the subjective probabilities of counterfactuals challenge not only standard semantics, but truth conditional accounts in general. It seems that under conditions of certainty, counterfactuals must at least be as strong as the standard truth conditions, whereas under conditions of uncertainty, they appear to express a weaker proposition. Taken together, it seems counterfactuals should and should not imply the standard truth conditions.

In analysing the puzzle generated in this way, we have seen that it relies on two background assumptions concerning the epistemic impact of implication relations, at least one of which seems problematic. If counterfactuals were weaker than the standard truth conditions but had a more demanding epistemic profile which would be such that they could

[14] In a similar vein, Edgington (2008) counters Bennett's objection by pointing out that she intends the argument to be run on the relevant hypothetical belief state rather than the actual belief state. So, by holding that counterfactuals behave like indicative conditionals in a hypothetical belief state, she intends to generalize the problem with the 'or'-to-'if' inference to counterfactuals.

only be known if one is in a position to know that the standard truth conditions obtain, then counterfactuals could have truth conditions while still satisfying the two assumptions of the puzzle.

This is admittedly only a small loophole in the argument against truth conditions. So before we explore how one might be able to exploit it, we shall discuss what might seem to be the more natural conclusion to be drawn from the puzzle. One may simply take the argument against truth conditions to be sound: the subjective probabilities of counterfactuals cannot be explained in terms of truth conditions. Rather, they should be given a non-standard explanation. Given the suppositional nature of the evaluation constraint, one may even think that looking for a propositional explanation was wrong-headed from the start. As we shall see, following up on this thought does not necessarily mean that counterfactuals do not have truth conditions (though this is an option), for setting up the puzzle presupposes that the subjective probabilities concern the counterfactual as a whole. If one can give an alternative explanation, one regains some territory to assign truth conditions to counterfactuals which would then not have to carry the burden of explaining the data about subjective probabilities of counterfactuals on which the puzzle is based. How all this might work will be the topic of the next chapter.

5

Restriction and Modification

The data concerning the subjective probabilities of counterfactuals make it hard if not impossible to account for them in terms of truth conditions. It is therefore a natural move to step back and ask whether they should really be seen as reflecting the likelihood of a proposition expressed by the counterfactual as a whole. One might even take the puzzle about truth conditions for counterfactuals as a reductio of this idea. But what could an alternative account look like?

On the linguistic level, we have treated epistemic expressions as if they take the whole counterfactual as a single argument. This reflects the assumption that the data concern the epistemic properties of a proposition expressed by the counterfactual as a whole. However, this is not the only way one can think of the interaction between operators and conditionals. Partly motivated by the problems of accounting for the similarly puzzling data concerning indicative conditionals, an alternative picture has been developed: the restrictor view of conditionals. Originating from a suggestion by Lewis (1975) and further developed in the works of Kratzer (see the collection of papers in Kratzer 2012) and others, the restrictor view requires a fairly radical departure from standard theories of conditionals by questioning the fundamental assumption that conditionals are to be seen as two-place connectives. The basic idea is that 'if'-clauses are restrictors of operators rather than connectives. On this picture, the semantics of 'if, then'-constructions is not given by assigning to 'if, then' a function which operates on two arguments, the semantic value of the antecedent and the semantic value of the consequent. Rather, the semantics of conditionals can only be given in relation to an operator the 'if'-clause can restrict. For this to be a general theory of conditionals, one needs to assume that there is always an operator present in the logical form of conditionals.

Although the restrictor view of conditionals is meant to apply both to indicative conditionals and to counterfactuals, it receives its main motivation from certain indicative constructions. In particular, the question of whether counterfactual 'if'-clauses can restrict epistemic modals expressing present epistemic modality has, as far as I am aware, not been addressed in any detail. For this reason, this chapter investigates whether the restrictor view can be put to use to explain the puzzling data concerning the subjective probabilities of counterfactuals. Given the potential of the restrictor view to explain how conditional probabilities could be expressed by indicative conditionals, there is reason to hope that it might yield an account of the subjective probabilities of counterfactuals, too, for they seem to flow in a similar way from a certain conception of conditional chances. As it turns out, the idea of a restrictor has to be generalized if such an explanation is to be had. What we will need is the idea of a *modifier* which can modify an operator in a way which is not restriction.

The chapter is structured as follows. After explicating the restrictor view in some more detail, we will see why the subjective probabilities of counterfactuals cannot be explained in terms of restriction. I will then introduce the idea of a modifier. As an interlude, I will briefly compare the resulting theory to two similar approaches: the suppositional view and a dynamic semantics of counterfactuals. We then turn to two potential problems for the view, a smaller and a bigger one. The latter is the main reason why I think it is worthwhile to continue to look for a propositional explanation of the problematic data.

5.1 The Restrictor View

To set the stage, here is a brief description of the restrictor view of conditionals. Lewis (1975: 10f.) developed the idea that 'if'-clauses might be capable of performing a restrictor function in order to explain the intuitive truth conditions of sentences with the following structure:

(1) Always/Usually/Sometimes, if it is raining, my roof leaks.

Lewis argues that 'if'-clauses occurring in sentences of this kind are best interpreted as restrictors of the adverbial quantifiers. The observation behind his argument is the following. The truth conditions for a construction with 'always' seem to correspond to a universally quantified material conditional:

(2) It is always the case that either it is not raining or my roof leaks,

but in order to get the truth conditions for 'sometimes' right, we would have to assume that the conditional works like a conjunction, for only then the sentence comes out true iff

(3) There is a time at which it is raining and my roof leaks,

which are the intuitive truth conditions. So, there is no single conditional meaning which is supplied both under 'always' and under 'sometimes'. And it would be implausible to assume that the embedded conditional clause varies its meaning depending on the quantifier under which it is embedded (but see Higginbotham 1986).

Even worse, for quantifiers like 'often' or 'usually', it seems virtually impossible to find a conditional meaning which would correspond to the intuitive truth conditions. The observation there is essentially the same as the argument for restricted quantifiers (cf. Barwise and Cooper 1981) in constructions like

(4) Most people enjoy watching TV.

The latter sentence is best analysed as a restricted quantification of the form

(5) $[\text{Most } x : Fx][Gx]$,

where the predicate '*F*' is taken to restrict the domain of the quantifier (square brackets here indicate the supposed syntactic phrase structure). Contrast this with an analysis like the following:

(6) $[\text{Most } x][Fx \rightarrow Gx]$,

where the quantifier would take a conditional as a single argument. For 'most' it is known that there is no (classical) connective which would provide the intuitive truth conditions on this syntactic analysis. Given the similarity in meaning between

(7) Often, if it is raining, my roof leaks

and

(8) Most cases in which it is raining are cases in which my roof leaks

it is not surprising that we face similar difficulties if we assume that 'often' quantifies over a whole conditional. The natural alternative would be to take the 'if'-clauses to be a restrictor of the quantifier in close analogy to cases of restricted quantification.

On a restrictor analysis, the interaction of 'if'-clauses with adverbial quantifiers could be given a unified account. The structure of Lewis's sample sentences would be something like the following, with Q representing an arbitrary quantifier:

(9) [Qt : it is raining at t][my roof leaks at t].

Thus, in the logical form of these sentences, we would not find a two-place conditional connective any longer. As Lewis puts it:

> I conclude that the *if* of our restrictive if-clauses should not be regarded as a sentential connective. It has no meaning apart from the adverb it restricts. (Lewis 1975: 11)

The proposal has natural extensions for other kinds of expressions. Most important in the present context is its extension to epistemic phrases expressing different degrees of likelihood. Here one could assume analogously—as Lewis (1975) already suggests—that an expression of probability might get restricted by the antecedent of the conditional. Consider:

(10) Probably, if I buy a lottery ticket, I will lose,

and the possibility of giving it a restrictor analysis:

(11) [Probably: I buy a lottery ticket] [I will lose].

The truth conditions seem very close in meaning to a clear case of restricted quantification:

(12) Most cases in which I buy a lottery ticket are cases in which I lose.

If the latter is given a restrictor analysis, it is a natural thought to consider the same analysis for the former, too. If the probability function associated with 'probably' is uniform, the two types of sentence can be taken to be equivalent.

To account for the general case in which 'probably' may not quite function like 'most' because it weighs some cases more than others, something has to be said about the effect of restricting an expression of probability. The first thing which springs to mind is to account for it in terms of conditionalization (cf. Kratzer 1986 and, for more details, Égré and Cosiz 2011). Given this assumption, it could be explained how embedding conditionals under probability operators results in expressions of conditional probability.

5.2 Simple Conditionals

Lewis does not make a universal claim about 'if'. What he says is compatible with the possibility that 'if' sometimes behaves like a two-place sentential connective. Kratzer has generalized Lewis's proposal.[1] She claims that 'if' is always a restrictor and never a sentential connective:

> The history of the conditional is the story of a syntactic mistake. There is no two-place *if... then* connective in the logical forms for natural languages. *If*-clauses are devices for restricting the domains of various operators. Whenever there is no explicit operator, we have to posit one. (Kratzer 1986: 656)

For many cases, Kratzer (1986) assumes that indicative conditionals are implicitly modalized by an epistemic necessity operator. Consider an utterance of a simple conditional which does not contain any overt operator the 'if'-clause might restrict:

(13) If Jacky arrived in New York yesterday, she contacted Jack.

On Kratzer's account, the logical form of what is said by this utterance typically looks like this:

(14) [MUST: Jacky arrived in New York yesterday] [she contacted Jack].

Utterances of indicative conditionals which are not explicitly modalized will typically express a restricted kind of epistemic necessity.[2]

[1] See Kratzer (1979), Kratzer (1981) (updated in Kratzer 2012: ch. 2) and perhaps most importantly Kratzer (1986) (updated in Kratzer 2012: ch. 4).

[2] There is no need to assume that the covert modal behaves like the overt modal 'must'. Without this assumption, one can, for instance, allow that indicative conditionals sometimes

Assigning truth conditions to simple, unmodalized counterfactuals is less straightforward, but the basic idea of taking them to restrict an implicit operator can still be brought to work.[3] Given that the truth conditions of counterfactuals do not generally coincide with the truth conditions of the corresponding indicative conditional, counterfactual 'if'-clauses would not be taken to restrict an epistemic necessity operator, but rather a different operator better suited to generate plausible truth conditions for counterfactuals. The difference between the two kinds of conditional could be explained by assuming that they typically select different implicit operators. What makes this project challenging is the fact that restrictable operators which could provide plausible truth conditions for counterfactuals are not so easy to find.

With some sophistication, however, it is possible to derive the standard truth conditions for counterfactuals (see Kratzer 1981).[4] To capture the variable strictness of the standard truth conditions, one would need to assume a more complex implicit operator. Informally speaking, the operator would have to express something like *close necessity*, so that the consequent only needs to be true at all the close or relevant antecedent-worlds. On this idea, the logical form of a counterfactual would roughly look like this:

(15) [CLOSELY NECESSARY: Antecedent] [Consequent].

Restricting an operator of close necessity to worlds at which the antecedent is true would generate a set of close antecedent-worlds. Saying that all of those are consequent-worlds would then be exactly what the standard truth conditions require.

have the material truth conditions by interpreting the necessity operator in a vacuous way. Cf. Kratzer (1986).

3 It would also be possible to hold that counterfactuals always contain an overt modal the 'if'-clause can restrict, namely the modal governing the consequent. In the case of simple counterfactuals, this is typically 'would'. If one wants to derive the standard truth conditions in this way, one could adopt the view that 'would' expresses something like close necessity as explained below. There is a further possibility which only becomes visible once the idea of modification, to be explained below, has been developed. Once in place, it would then be possible to hold that counterfactual 'if'-clauses do not restrict a covert (or overt) operator, but rather modify it.

4 An alternative picture is developed by von Fintel (2001), who takes counterfactuals to be strict conditionals with a contextually restricted domain, such that counterfactuals will effectively often have truth conditions very similar to the standard truth conditions.

It will help to see how this works in detail. The basic idea is that a modal can, in addition to a world of evaluation w and a domain of worlds D_w, be sensitive to an ordering source $\leq_w$.[5] An ordering source tells us which worlds are closer than others.

Modelled after Lewis's (1973: 16) specification of truth conditions for counterfactuals, Kratzer defines a notion of necessity which is relative to an ordering of worlds in the following way:

Necessity (Kratzer). A proposition p (expressed by α) is a *necessity* in w with respect to f and g ($[\![\text{NECESSARY } \alpha]\!]^{w, D_w, \leq_w} = 1$) iff for all $u \in D_w$, there is a $v \in D_w$ such that
(i) $v \leq_w u$
and
(ii) for all $z \in D_w$: if $z \leq_w v$, then $z \in p$.

So, roughly speaking, a proposition is said to be a necessity if it obtains at all $\leq_w$-closest worlds in the domain D_w. Since Kratzer wants to allow for worlds getting closer and closer without there being some closest worlds—thus denying the limit assumption (see section 7.4)—the definition has to be a little more complex. For a proposition to be necessary, it is required that there is to any world u in the domain a $\leq_w$-closer world v such that the proposition holds at all worlds $\leq_w$-closer than v. In other words, if we approach w along the ordering relation, we find from a certain point on only p-worlds.

For specific settings of the domain and the ordering parameter, certain familiar kinds of necessity result. If $\leq_w$ relates all worlds, then the impact of the ordering source vanishes and necessity converges to ordinary necessity over the domain D_w. If we further let D_w be the set of all metaphysically possible worlds, we see that broad metaphysical necessity turns out to be a limiting case. The implicit modal in the restrictor view's analysis of counterfactuals, however, will require a non-trivial ordering as employed in standard semantics for counterfactuals. A necessity operator of this kind would express something like close necessity.

[5] I simplify here somewhat the proposal in Kratzer (1981). Instead of a set of worlds as the domain, Kratzer works with a richer notion of a *conversational background* which is construed as a set of propositions, alias sets of worlds. But in the semantics for modals which interests us, only their intersection matters, which is simply a set of worlds. Similarly, she has a richer notion of an ordering source, but in the semantics only the ordering relation induced by it plays a role.

This gives us already a fairly general theory of conditionals. They are a device for restricting various operators or quantifiers. In unembedded contexts, they typically express a strict epistemic conditional in the indicative case and the standard truth conditions in the counterfactual case by restricting an implicit necessity operator. Embedded under an overt operator or quantifier, they restrict the corresponding domain to those cases at which the antecedent holds.

It is important to note that giving a semantics of counterfactuals in this way is not challenged by the puzzle about their truth conditions, for the subjective probabilities are, at least as far as they are reflected in the linguistic behaviour of epistemic expressions, no longer construed as being concerned with the probability that these truth conditions obtain (I will comment on how this extends to credences in counterfactuals in section 5.5). Saying

(16) It is likely that if I had bought a lottery ticket, I would have lost

is not taken to attribute high probability to the proposition the counterfactual expresses when it occurs unembedded. Actually, it is not taken to attribute probability to the counterfactual at all. Rather, it is taken to attribute probability to the consequent relative to a domain of possibilities which has been restricted by the antecedent. Thus, the restrictor view promises to explain the data about the subjective probabilities of counterfactuals without having to give up standard semantics for counterfactuals when they occur unembedded.

5.3 Counterfactual 'If'-Clauses as Restrictors

As it stands, the restrictor view of counterfactuals cannot be quite adequate. As we shall see in this section, it would assimilate counterfactuals to indicative conditionals in contexts in which they are both embedded under an epistemic expression. To see the problem, let us look at the classic Oswald/Kennedy example (see Adams 1970) embedded under a probability operator:

(17) *It is likely* that no one else would have killed Kennedy if Oswald had not done it.

Assuming here that the epistemic modal is restricted by the counterfactual 'if'-clause would yield an interpretation according to which most present epistemic possibilities in which Oswald did not kill Kennedy are such that no one else killed Kennedy.[6] But this is an undesirable consequence: In every present epistemic possibility someone killed Kennedy, for it is common knowledge that someone killed Kennedy. Restricting the epistemic possibilities to those in which Oswald did not do it will not alter this fact. In all of those, it will still be the case that someone else killed Kennedy, since they are merely a subclass of our original class. As a consequence, (17) would come out as false. But it seems to be true.

In fact, the present analysis would have the following statement come out true:

(18) *It is certain* that someone else would have killed Kennedy if Oswald had not done it.

This would turn out to be true, for Kennedy is killed by someone else at all worlds which are presently live possibilities. So, when 'it is certain' is evaluated with respect to this domain and applied to the proposition that someone else killed Kennedy, then the resulting expression comes out true. But, of course, the present statement should come out false: not only is the embedded counterfactual not certain to be true, it is likely to be false.

The following may lead to a diagnosis of the problem. Applying the restrictor analysis to the interaction between epistemic modals and counterfactual 'if'-clauses would assimilate (17) to its indicative version:

(19) *It is likely* that no one else killed Kennedy if Oswald did not do it.

This ascription of epistemic probability is indeed incorrect and its incorrectness may be explained by assuming the 'if'-clause to restrict the probability operator, for at no present epistemic possibility at which Oswald did not kill Kennedy is it the case that no one else did.

Assuming that counterfactual 'if'-clauses restrict epistemic modals in the same way as indicative 'if'-clauses are supposed to do does not leave room for the differences between the two kinds of conditionals

[6] To simplify, I assume here that the relevant epistemic possibilities are equally likely. Otherwise 'it is likely that' would not necessarily go by what is the case at most epistemic possibilities.

when they are embedded under epistemic operators. The same probability ascriptions would always be true of both kinds of conditional. So, we are driven to the conclusion that the semantic relation between epistemic modals and counterfactuals cannot be accounted for on the model of domain restriction.

The problem with the restrictor analysis seems to be that it treats counterfactual 'if'-clauses just like indicative 'if'-clauses when they interact with a modal operator. But Oswald/Kennedy type examples suggest that at least with respect to epistemic modals, the behaviour of counterfactual 'if'-clauses and indicative 'if'-clauses comes apart. This happens in very much the same cases in which an unembedded indicative conditional is evaluated differently from the corresponding counterfactual. Explaining the difference in the unembedded case is not a problem, for, as we have seen, it may be assumed that different implicit operators are selected. But once the operator is explicit and the same in both cases, there is no room anymore for a possible difference.[7]

One assumption I have made is that epistemic phrases such as 'it is likely' still express present epistemic possibility under the salient interpretation when they attach to counterfactuals. One way of seeing that this is indeed the case is to focus on cases where the probability ascription is coordinated with an ascription of probability to an ordinary proposition. For instance, consider

(20) *It is likely* that Oswald shot Kennedy and that no one else would have done so if Oswald had not done it.

Given that it is clear that we want to express high present epistemic probability concerning a past event as far as the first conjunct is concerned, the probability ascription to the second conjunct must be given the same interpretation.

[7] There is one possibility I have ignored here: one could assume that in the counterfactual case the 'if'-clause continues to restrict an implicit operator which scopes under the probability operator (for general discussion of doubly modalized conditionals, see e.g. Geurts 2004, Kratzer 2012: 106f., and Swanson 2010). But then the attributions of probability to counterfactuals would be explained as attributions to a proposition expressed by the counterfactual as a whole just as if we had fed in the standard view. However, because of the problems with this possibility, we are currently looking for a solution in which the restrictor view could do some essential work.

Alternatively, one could also consider epistemic attitude verbs as in the following example:

(21) *Jacky is fairly confident* that no one else would have killed Kennedy if Oswald had not done it.

As present tense epistemic attitude verbs do not have interpretations on which they do not go by the present epistemic attitudes of the subject, there is no room for manoeuvre here for arguing that embedding counterfactuals under such expressions might make another interpretation salient which no longer concerns the present epistemic possibilities.

Another assumption I have made is that the epistemic phrases under consideration are effectively only sensitive to a domain of discourse and not to an ordering source. This is a standard assumption. Still, one may wonder whether one could get around the present problem by allowing the ordering source to play a role as well. However, I cannot see how this might be possible. The problem seems to be due to a structural feature which cannot be outmanoeuvred by bringing an ordering source into play. When we evaluate a counterfactual, we often consider worlds which are not live (epistemic) possibilities. In the Oswald/Kennedy example, we consider worlds at which no one killed Kennedy although it is not a live possibility that no one killed Kennedy. So, the restricted domain of discourse with respect to which the epistemic phrase is evaluated should at least sometimes contain worlds which are not contained in the original domain to which the epistemic modal is sensitive. But this effect cannot be achieved by an ordering, for independently of its effects, the restricted domain will always be a subset of the original domain. What we would need is a partial expansion of this domain. This observation gives rise to what might be a better proposal: counterfactual 'if'-clauses do not simply function as restrictors, they interact with epistemic domains in a more complex way by partially expanding them. To this idea we now turn.

5.4 Counterfactual 'If'-Clauses as Modifiers

Epistemic phrases interact with counterfactual 'if'-clauses in a way which is not restriction. If one thinks about it, this may not be much of a surprise. After all, we evaluate indicative conditionals and counterfactuals in quite different ways. It was to be expected that this difference would be preserved when they are embedded under an epistemic modal. So, if we wish to stick to the basic idea of the restrictor view, we should allow counterfactual 'if'-clauses to interact with epistemic phrases in a similar but different way.

At this point, it might be helpful to go back to our discussion of indicative versus counterfactual suppositions in chapter 3. Indicative and counterfactual 'if'-clauses seem to invite us to do different things. An indicative conditional is evaluated by temporarily adding the antecedent to our stock of beliefs and then asking whether the consequent is true in this more informed belief state. There is a natural connection with the restrictor view here. Adding a proposition to one's stock of belief is like restricting one's belief state to those cases in which the proposition is true. If a belief state is modelled somewhat coarse-grainedly as a set of epistemically possible worlds, then adding a proposition conceived of as a set of worlds amounts to the intersection of the two sets or, equivalently, to the restriction of the original domain to the subset in which the proposition is true. Moreover, the fact that making an indicative supposition goes by conditionalization corresponds on the linguistic level to the fact that restricting an expression of probability results in an expression of conditional probability.

Counterfactual 'if'-clauses also invite us to make a supposition, but a counterfactual supposition is different from an indicative one. Making a counterfactual supposition often requires us to temporarily suspend certain beliefs while holding others constant. For instance, when we make the counterfactual supposition that Oswald had not killed Kennedy, we suspend our present belief that Kennedy was killed by someone but we hold onto our present beliefs about the circumstances a short while before Kennedy's assassination. In contrast, when we make the indicative supposition that Oswald did not kill Kennedy, we do not suspend our belief that Kennedy was killed by someone. As we have seen, where the effect of an indicative supposition may be described in terms of conditionalization, the effect of a counterfactual supposition could be described in terms of imaging. Integrating a counterfactual supposition takes us to the image of our present belief state under this supposition. Taking this to the linguistic level, the idea would be that counterfactual 'if'-clauses do not restrict epistemic phrases but sometimes perform a linguistic analogue of imaging. To have a name for this, let us say that counterfactual 'if'-clauses might *modify* the domain of epistemic expressions.

To get an idea of how this might work formally, let us start by formalizing the semantics of the restrictor view. At its centre, the restrictor view contains the idea that 'if'-clauses restrict the domain of the modal in the scope of which they occur. Syntactically, the restrictor view requires that the 'if'-clause combines with the modal to form a more complex modal rather than that it first interacts with the consequent to form a conditional sentence which would then be the single argument of the modal. Once we assume a syntactic analysis which gives us this result, the semantics only needs a clause which deals with the interaction of an 'if'-clause with a modal. To keep things simple, let us model the domain of modals as sets of possible worlds and let us ignore other parameters they might be sensitive to. Then the semantic clause for restriction by an 'if'-clause might simply look like this (cf. Kratzer 2012: 94):

$$(22)\quad [\![\mathrm{MODAL} : \mathrm{A}]\!]^{E} = [\![\mathrm{MODAL}]\!]^{E \cap [\![A]\!]}.$$

The semantic value of a modal restricted by an 'if'-clause A relative to an epistemic domain E is the semantic value of the modal relative to the domain E intersected by the semantic value of A. This clause gives us the desired result that the 'if'-clause functions as a restrictor on the domain of the modal. Of course, to derive the semantic value of the restricted modal we need a clause for the modal expression relative to a parameter and, quite obviously, the semantic value of A. But the latter two values will be derivable independently of what we say about conditionals.

Given a semantic clause along these lines, its application to complete conditional constructions including a consequent is straightforward. If we have a conditional governed by a modal which is syntactically analysed according to the restrictor view, the semantic clause dealing with restriction does all the interesting work:

$$(23)\quad [\![[\mathrm{MODAL} : \mathrm{A}][B]]\!]^{E} = [\![\mathrm{MODAL}]\!]^{E \cap [\![A]\!]}[\![B]\!].$$

The antecedent of the conditional restricts the domain of the modal by intersecting it. The only further thing which happens is that the restricted modal takes the consequent as an argument.

To give the semantics of the envisaged view according to which counterfactual 'if'-clauses can sometimes act as modifiers rather than restrictors, we can keep the structure of the semantics of the restrictor view. What would have to change is the operation of restriction: instead of accounting for it in terms of intersection, we would have to postulate a more complex operation F which describes how a counterfactual supposition A gets integrated into an epistemic state E yielding a new belief state $F(E, A)$. To identify the general structure, it might for the moment be better to leave it at an abstract level, but we shall soon see that we can account for modification in terms of imaging. In addition, it helps to assume that the operation of modification is visible on the syntactic level. I will write '::' if we are dealing with modification instead of restriction. Then the relevant semantic clause which accounts for the interaction between a modal and a counterfactual 'if'-clause could look like this:

(24) $[\![\text{MODAL} :: \text{A}]\!]^E = [\![\text{MODAL}]\!]^{F(E,[\![A]\!])}$.

The semantic value of a modal modified by a counterfactual 'if'-clause A relative to a domain E is the semantic value of the modal evaluated relative to the domain modified by the semantic value of A.

Now, in our discussion of imaging, we have focussed on how to define the image of a probability function. In order to explicate the desired operation of domain modification, we have as input not a probability function but only a domain consisting of a set of possible worlds.[8] But this actually makes our life easier, not harder. The difficult task in defining imaging was to explain the probability shift. Explaining how the domain changes is straightforward. Recall that the basic idea of imaging was to consider for every epistemic possibility w the set $f_w(A)$ of relevant antecedent-worlds. If we do this for all epistemic possibilities, we arrive at the domain the image is defined on. These are all the worlds w' for which there is some epistemic possibility w in our original domain such that w' is in $f_w(A)$, i.e. is a relevant antecedent-world relative to w. Our modified domain of epistemic possibilities would thus be the set of worlds for which it currently is an epistemic possibility that they are a relevant antecedent-world. So, we can define the modified domain $F(E, A)$ of a domain of epistemic possibilities E and the content A of a counterfactual 'if'-clause as follows:

(*) $F(E, A) := \bigcup_{w \in E} f(w, A)$.

[8] Some epistemic expressions, most notably expressions of probability, may not (only) be interpreted relative to a domain but relative to a probability function. In such cases, imaging as defined in the previous chapter would already fit the required structure.

Here we represent the modified domain as resulting from summing up all the sets of antecedent-worlds which are relevant relative to at least one world we take to be epistemically possible (in a dynamic setting, Starr 2014: Sect. 3.3 uses this approach to define expansions of domains; cf. also Veltman 2005).

To complete the account, let us now also state the clause for how a modal expression could get modified in the context of a counterfactual:

(25) $[\![[\text{MODAL} :: \text{A}][B]]\!]^{E} = [\![\text{MODAL}]\!]^{F(E,[\![A]\!])}[\![B]\!].$

Given that the modal's initial domain is E, the counterfactual antecedent's semantic value would modify this domain to $F(E, [\![A]\!])$. Relative to this modified domain, the modal would then be evaluated by feeding in the semantic value of the consequent as its main argument. Call this way of explaining the interaction of counterfactuals with epistemic modals the *modifier view*.

To see the explanatory potential of this approach, let us sketch how it resolves the problem posed by the Oswald/Kennedy case. Recall the two sample sentences

(26) It is likely that if Oswald did not kill Kennedy, someone else did.

(27) It is likely that if Oswald had not killed Kennedy, someone else would have.

The problem is that the first ascription of probability is fine whereas the second is not warranted. If we were to assume that the epistemic phrase is restricted in both cases by the 'if'-clause, we would not be able to mark this difference. This problem disappears once we assume that the epistemic phrase is restricted only in the first case but modified in the second one. For the counterfactual to be likely, it would need to be likely that someone else would have killed Kennedy from an epistemic view point obtained by counterfactually supposing that Oswald had not killed Kennedy. But from this perspective, someone killing Kennedy is not likely. Modification is capable of reflecting this way of evaluating counterfactual thoughts by allowing partial expansions of our domain of epistemic possibilities.

5.5 Variants of the View

The restrictor view and its extension, the modifier view, is just one representative of a family of views which provide non-propositional explanations of the data about the subjective probabilities of conditionals. There are at least two further approaches which deserve mentioning: *dynamic semantics* and *the suppositional view* of conditionals. In this section, I will briefly explain why I think that these two views can to a large extent be seen as variants of the restrictor view. I also think that they face similar problems (to which we will turn in the next two sections), but I will not attempt to argue for this further claim.

Start with dynamic semantics of conditionals. In a dynamic framework, the basic notion is that of a *context-change potential* of a sentence. It is designed to capture the effect an affirmative utterance of that sentence would have on a given context. A context can be seen as a state of information s represented by a set of possible worlds. The semantic role of a sentence ϕ is taken to be a function from contexts to contexts describing how the sentence changes a context. The update of a state s with a sentence ϕ is typically written as $s[\phi]$. Updates can be *classical* in the sense that the update $s[\phi]$ is always the intersection of s with a set of worlds coming from a classical interpretation $[\![\phi]\!]$ of ϕ, i.e. for all s, $s[\phi] = s \cap [\![\phi]\!]$. But not all sentences need to have a classical context-change potential. A state s is said to *accept* a sentence ϕ, formally $s \models \phi$, iff the update of s with ϕ does not change s, i.e. $s[\phi] = s$ (cf. Veltman 1996). Acceptance means that the information provided by ϕ is already contained in s. If the update by ϕ is classical, this means that s is contained in $[\![\phi]\!]$, i.e. s seen as a proposition implies $[\![\phi]\!]$.

In a dynamic framework, the semantics of conditionals is very similar to the restrictor view: the antecedent is taken to temporarily update the context with the antecedent and it is then checked whether the consequent is accepted in the updated context (cf. Gillies 2004, 2009). So, a conditional 'if ϕ, ψ' is accepted in a context s iff the consequent is accepted in the context resulting from updating the original context with the information contained in the antecedent, i.e. iff $s[\phi] \models \psi$. Effectively, the antecedent is taken to restrict the original context and it is then checked whether the consequent is accepted in the restricted context (in a non-dynamic setting, this semantics corresponds to a strict epistemic conditional, i.e. to a material conditional governed by an epistemic necessity operator).

A dynamic semantics for counterfactuals can be given along the lines of the modifier view. Here one would say that a counterfactual is accepted in a context iff the consequent is accepted in the context resulting from the original context by modifying it with the information of the antecedent (cf. Starr 2014, Veltman 2005; see also von Fintel 2001). There are at least two interesting differences from the restrictor view as sketched in the previous sections. One is that a dynamic semantics of conditionals can be supplemented with a genuinely dynamic notion of consequence (see Veltman 1996: sec. 1.2 for discussion of various ways of doing so). Another is that modals attaching to conditionals are typically reconstructed as having narrow scope over the consequent only. The antecedent of the conditional is first taken to create a temporarily updated context relative to which the modalized consequent can then be evaluated. But the ultimate effect on the interpretation of the modal is the same. Both on the domain restrictor view and on the dynamic view the modal is interpreted relative to the restriction of the original domain by the information contained in the antecedent.

There is also a close connection between the restrictor view and the supposition view of conditionals.[9] On the suppositional view, the core function of an 'if'-clause is to express a supposition in the scope of which the consequent is to be evaluated. Epistemic attitudes towards a conditional are not categorical attitudes towards a proposition expressed by the counterfactual as a whole. Rather, they are conditional attitudes towards the consequent of the conditional relative to the supposition of the antecedent. For instance, having a high credence towards a conditional is

[9] The suppositional view has already a fairly long history. Quine (1950) mentions the suppositional view and attributes it to Philip Rhinelander. Von Wright (1957: ch. 7) may contain the first systematic development of it. Other early proponents of the suppositional view include Adams (1965, 1975), Dummett (1973, 1981: 338–52) and Belnap (1970). More recent proponents include Barnett (2006) and Edgington (1986, 1995). Certain authors like Bennett (2003) and Gibbard (1981) accept the suppositional view for indicative conditionals, but reject it for counterfactuals. A special case is Woods (1997), who accepts the suppositional view only for indicative conditionals which do not have 'will' as the main auxiliary verb in the consequent. Until recently, there was a stronger focus on indicative conditionals, though Adams (1975) has already a chapter on counterfactuals. But over the last years Barnett (2010) and Edgington (2008) have discussed the counterfactual case in more detail.

explained as having a high conditional credence in the consequent under the assumption of the antecedent. The suppositional view offers a parallel story for assertions of conditionals and other kinds of speech-acts. Asserting a conditional is not taken as a categorical assertion, but rather as an assertion of the consequent relative to the supposition of the antecedent.

As we have seen, restriction and modification are linguistic analogues of indicative and counterfactual suppositions. Restricting the domain of an epistemic expression has the same effect on the domain as making an indicative supposition would have on a belief state which is adequately represented by that domain. The same holds for the relation between modification and a counterfactual supposition. The suppositional view can therefore be seen as a non-linguistic variant of the restrictor view.

The suppositional view is compatible with how the restrictor view explains embeddings of conditionals under restrictable operators. Actually, this is probably the most natural way proponents of the suppositional view can account for such embeddings. The difference between the two views shows up in how they account for unembedded conditionals. While the restrictor view posits an implicit operator for the 'if'-clause to restrict, proponents of the suppositional view typically hold that conditionals do not have (ordinary) truth conditions. However, it is not hard to see how the semantics of the restrictor view could be modified to yield this result. One could simply not posit an implicit operator for unembedded conditionals. In the interpretation of simple conditionals, the operator slot would be left open (we may call such interpretations *open interpretations*). Simple conditionals would then not have truth conditions. They would only express a conditional construction whose modal force is left unspecified. It would, of course, be a further question whether one could derive all the properties the supposition view attributes to conditionals from open interpretations of conditionals.

One advantage of the suppositional view is that it allows for an explanation of the subjective probabilities of conditionals. As we have noticed in passing, the restrictor view merely explains how conditionals embed under epistemic phrases which might express possibility, high probability or certainty. So, the restrictor view, or rather its extension in the form of the modifier view, has a story about why the complex construction

(28) It is likely that if I had bought a lottery ticket, I would have lost

can be true. It is not designed to explain how it is that we are *fairly certain* about the embedded counterfactual when it occurs unembedded:

(29) If I had bought a lottery ticket, I would have lost.

It is clear that what we are not fairly certain about is that the standard truth conditions obtain. So, it is not an option for the restrictor view to account for our credences by construing them as credences in a proposition which would be expressed by the counterfactual if we took it to be implicitly modalized by a necessity operator so that it would have the standard truth conditions. In contrast, the suppositional view is designed to have an explanation for this type of case. Our credences are not construed as ordinary credences in a proposition; they are rather taken to be a specific kind of conditional credence attaching to the consequent in the scope of counterfactually supposing the antecedent. If the restrictor view could adopt this story by saying that our intuitions about credences derive from open interpretations of conditionals, it would gain significantly in explanatory power.

The absence of truth conditions on the suppositional view becomes a problem when conditionals are embedded in contexts where the 'if'-clause cannot restrict the embedding phrase, say when the conditional occurs as a conjunct in a larger construction (see the exchange between Edgington 2000 and Kölbel 2000). Here it faces a variant of the Frege-Geach problem. Taking conditionals to be implicitly modalized in such constructions as it is assumed on the restrictor view would avoid this kind of problem, although we will see below that there may still be a more specific problem of embedding for the restrictor view, too.

5.6 The Lack of Independent Evidence

On the restrictor view, indicative conditionals interact with a variety of quantifiers and other operators in the same way as they interact with epistemic modals. Thus, the interaction between epistemic modals and indicative 'if'-clauses is not described as an isolated phenomenon, but rather as an instance of a much more general pattern. Given this, one would expect that the operation of modification which can be employed to explain the interaction of epistemic modals with counterfactual 'if'-clauses will show up at other occasions as well. Somewhat surprisingly, this does not seem to be the case. As far as I can see, the modifier view lacks independent evidence. Let us look at some relevant cases.

Start with the classical case of adverbs of quantification employed in Lewis 1975 to highlight the restrictive function of 'if'-clauses. The problem here is that adverbs such as 'sometimes', 'always', 'often', or 'usually' do not happily combine with counterfactuals. Compare

(30) Sometimes/Always/Often/Usually, if I buy a lottery ticket, I win

with

(31) ? Sometimes/always/often/usually, if I had bought a lottery ticket, I would have won

and

(32) ? Sometimes/always/often/usually, if I were to buy a lottery ticket, I would win.

In the indicative case, the 'if'-clause can be interpreted as restricting the adverbial quantifiers. Substituting the indicative clause with its counterfactual variant, either past or present, yields sentences which are hard to interpret.[10] The counterfactuals may not be completely uninterpretable, though. With a bit of a stretch, it might be possible to hear them as saying, for instance, that there are certain times *t*, such that if I had bought a ticket at *t*, I would have won. But note that if such interpretations are available, the counterfactual 'if'-clause does not seem to restrict, let alone modify the adverbial quantifier. The constructions used to motivate the restrictor view in the indicative case cannot be used to motivate the modifier view in the counterfactual case.

In the indicative case, there is some evidence that 'if'-clauses might be able to restrict nominal quantifiers. Consider, for instance,

(33) Most letters are answered if they are shorter than 5 pages.

(34) Few people like New York if they didn't grow up there.[11]

[10] That these counterfactuals are problematic was brought to my attention by Manfred Krifka in discussion after a talk in Berlin in 2010. I used to think that these constructions are strictly uninterpretable. Various participants at a workshop on conditionals in Konstanz in 2013 persuaded me that this may not be the case.

[11] These examples are given in von Fintel and Iatridou (2002) where they are attributed to Heim and/or Kratzer.

It seems quite natural to hear these sentences as equivalent to variants of them in which a relative clause restricts the quantifier phrase:

(35) Most letters which are shorter than 5 pages are answered.

(36) Few people who didn't grow up there like New York.

Contrast the indicative constructions with their counterfactual counterparts:

(37) Most letters would have been answered if they had been shorter than 5 pages.

(38) Few people would have liked New York if they had not grown up there.

In neither case does the counterfactual 'if'-clause modify the domain of quantification. Nor does it seem to restrict the domain in a way analogous to the indicative case. Where (35) has an interpretation on which the domain consists of the letters shorter than 5 pages, the domain in (37), (35)'s counterfactual variant, seems to consist of all (contextually salient) letters, not somehow of the letters satisfying the antecedent of the counterfactual. Rather, most of all letters are said to have the complex counterfactual property of having been answered if they had been shorter than 5 pages.

In the literature about the interaction between conditionals and deontic modals, there is evidence that indicative conditionals with a deontic modal governing their consequent sometimes receive an interpretation according to which the 'if'-clause is taken to be a restrictor. Consider, for example,

(39) If Beryl is smoking marijuana, we should report her to the police.[12]

There seems to be an interpretation of (39) according to which it expresses a conditional obligation rather than an unconditional obligation made on the assumption of Beryl smoking marijuana.

The possibility of such an interpretation is clearly absent in the counterfactual case:

[12] This example is due to Geurts (2004).

(40) If Beryl had been smoking marijuana, we should have reported her to the police.

More importantly even, we do not get a reading on which the counterfactual 'if'-clause would somehow modify the deontic modal.[13]

So far, we have seen pairs of cases in which the counterfactual variants could neither be interpreted as modifiers nor as restrictors. Under certain constructions of objective modality, we seem, however, to find cases in which counterfactual 'if'-clauses act as restrictors. Consider:

(41) There was a high chance that Chelsea would have won if the referee had not denied them a penalty.

If one finds that this example is an expression of past high conditional objective chance—which is an intuitively plausible verdict—then we could account for this by assuming that the counterfactual 'if'-clause restricts the chance operator. It is important to note that in order to get this interpretation, we would only need the counterfactual 'if'-clause to be a restrictor and not a modifier, for simple conditionalization is achieved through the restriction of a probability function, not somehow through a modification of its domain.

All in all, it seems that one cannot gather independent evidence for the modifier view. The way counterfactual 'if'-clauses interact with certain epistemic phrases would be a fairly isolated phenomenon. How much of an objection is this against the modifier view? Perhaps the first thing to note is that the modifier view derives a further bit of evidence from the fact that it is a natural extension of the restrictor view. Jointly, the two views form a homogeneous package which provides a highly unified account of the two sets of data. Both in the indicative and in the counterfactual case, the puzzling constructions are explained by challenging the logical form usually attributed to them. But given that the modifier view of counterfactuals is based on a much smaller amount of evidence, there is clearly room for thinking about an alternative explanation of the relevant data. We shall now turn to what strikes me to be a deeper problem for the modifier view (and perhaps for the restrictor view in general).

[13] 'Should' can also be given an epistemic interpretation as might be salient in 'If Beryl had been smoking marijuana, we should have noticed'. The claim made in the text is meant to concern deontic interpretations only.

5.7 The Problem of Embeddings

There is a certain class of embedded occurrences of counterfactuals for which the restrictor view is designed to have an explanation. These are embeddings under various operators, in particular epistemic operators, which can be restricted or modified by an 'if'-clause. As we shall see, there are other, fairly ordinary contexts like conjunction or existential generalization, where restriction or modification is not (always) a plausible option. It is at this point where the possibility of assigning truth conditions to conditionals by positing an implicit operator becomes a valuable tool, for one may assume that in such contexts conditionals contribute the truth conditions which would result from their implicit modalization.

Although this suggests that the restrictor view has a ready-made story for embedded conditionals—the 'if'-clause either restricts an overt or a covert operator—I will now present an argument which suggests that the problem cannot always be resolved in this way. I will focus here on counterfactuals only, but the structure of the problem should make us strongly suspect that there is a corresponding problem in the indicative case. What I will try to show is that the present solution leads to a certain kind of unwanted *transition failure*. If counterfactuals play a vastly different role when they occur in the scope of certain epistemic operators and when they are embedded in other contexts, then certain transitions between these two kinds of context will not be possible. Intuitively, however, these transitions seem to be possible as we shall see.

Let us start by considering the case in which the standard truth conditions are assigned to counterfactuals because the counterfactual 'if'-clause does not modify an overt operator. As will be obvious from the discussion to follow, our considerations will equally apply to close relatives of the standard truth conditions and in fact to many other possible truth conditions as well. But for ease of exposition, and since it is the most interesting case anyway, I will describe the problem with respect to the standard truth conditions.

There are two important kinds of context with respect to which we can expect that counterfactuals would need to be assigned truth conditions. We have seen in the previous section that ordinary objectual quantifiers are not modifiable by counterfactual 'if'-clauses. With respect to certain truth-functional connectives like conjunction, it does not even seem to be an option that they are somehow modified by counterfactual 'if'-clauses. So, these expressions generate contexts for which the present strategy of explaining embeddings of counterfactuals in terms of truth conditions is likely to be employed.

To illustrate these two kinds of context, I will focus on existential quantification and conjunction as in the following two examples:

(42) There is someone who would have got angry if I had criticized him.[14]

(43) There is blood on the kitchen knife and if the cook had done it, there would have been blood on the kitchen knife.

On the present suggestion, the embedded (open) counterfactual in the first example has the standard truth conditions within the scope of the existential quantifier. So, the sentence would be true just in case there is someone who gets angry at all relevant worlds at which I criticize him. Similarly, the counterfactual in the conjunction would contribute the standard truth conditions. The conjunction would then be true just in case there is blood on the kitchen knife and at all relevant worlds at which the cook commits the murder, there is blood on the kitchen knife.

Now, to begin with, compare the counterfactual in the first example to the following occurrence of a closed variant of it within the scope of an epistemic operator:

(44) It is quite likely that if I had criticized Bobby, he would have got angry.

For the sake of argument, suppose that this is a case in which the counterfactual 'if'-clause 'if I had criticized Bobby' modifies the epistemic operator 'it is quite likely'. According to the modifier view, the present example can be true if we take most but not all relevant worlds at which Bobby is criticized to be worlds at which he gets angry. Thus, for the example to be true, it does not have to be the case that we have sufficient information to be justified in thinking it likely that the standard truth conditions are satisfied.

[14] The choice of this example is inspired by the discussion of embeddings in Kölbel (2000).

The counterfactual in the present example comes from the open counterfactual used in example (42) above by substituting 'Bobby' for the quantified variable. Nevertheless, according to the modifier view, the two counterfactual 'if'-clauses perform quite different roles in the two constructions. Within the environment of the existential quantifier, the counterfactual would contribute the standard truth conditions because the counterfactual 'if'-clause will be taken to restrict or modify an implicit necessity operator (or alternatively the overt 'would'; cf. fn. 3 of this chapter). In the scope of the epistemic modal, however, the counterfactual 'if'-clause is taken to modify the epistemic operator. Modifying the epistemic operator will not be equivalent to taking the 'if'-clause to first modify a covert necessity operator and then to apply the epistemic modal to the resulting sentence, for the resulting sentence will have the standard truth conditions and the interaction of epistemic modals and counterfactuals cannot be explained in terms of the standard truth conditions.

Now, if such different roles are assumed, then certain transitions between the different sentential contexts are not possible. To see this, let us add the following example:

(45) It is quite likely that there is someone who would have got angry if I had criticized him.

This example results from the original existential quantification (sample sentence (42)) by applying the epistemic operator 'it is quite likely' to it. Now, intuitively, the present example is acceptable on the grounds of example (44). If we think it quite likely that Bobby would have got angry if I had criticized him, then we seem to be entitled to think it quite likely that someone would have got angry if I had criticized him. The reason for this is simple: existential generalization remains to be valid when it comes to counterfactuals. So, from

(46) If I had criticized Bobby, he would have got angry

the sample sentence

(42) There is someone who would have got angry if I had criticized him

follows logically. The problem is that we seem to be justified in performing this inference even within the scope of a probability operator.

However, the transition from (44) to (45) does not come out as justified on the modifier view. Since the counterfactual 'if'-clause in (45) occurs within the scope of the existential quantifier (and is an open sentence anyway), it cannot be taken to modify the outer epistemic operator. As a consequence, the quantified counterfactual contributes in (45) the same proposition as (42) expresses in its standalone use. So, it will be true if it is likely that there is someone who is angry at all relevant worlds at which he is criticized. But taking this proposition to be likely is not always justified on the grounds of having a high relative credence that Bobby would have got angry on the supposition that I had criticized him. Having this modified epistemic attitude does not require having a high credence that Bobby gets angry at *all* relevant worlds at which he is criticized. So, if Bobby is in the present context the only candidate for being criticized, then it is clear that having this modified epistemic attitude should not incline us to think it likely that there is someone who gets angry at *all* relevant worlds at which he is criticized. The phenomenon of unwanted transition failures indicates that the modifier view cannot adequately deal with embedded counterfactuals as exemplified by (45).

To clarify the problem a little further, let us look at it from a slightly different angle. The sample sentences we are considering have the following surface structure:

(47) $\Delta(A(t) > B(t))$.

(48) $\Delta(\exists x(A(x) > B(x)))$.

In example (44), a probability operator is applied to a counterfactual whose antecedent and consequent contain a singular term. And sample sentence (45) results from this example by existentially generalizing into the position of the singular term from within the scope of the probability operator.

The way we use these sentences suggests that we take their surface structure to be genuine. By accepting a sentence of the first form, we feel entitled to accept a sentence of the second form based on the idea that if a certain proposition is likely, then one of its logical consequences will be likely as well. This is what is reflected in the fact that if we take a counterfactual to be likely, we feel entitled to accept an existential generalization on it to be likely, too.

Now, the modifier view has it that the two sample sentences have quite different logical forms, and, most importantly, the logical role of the counterfactual 'if'-clause is taken to vary significantly in the two examples. To a good approximation, we may represent the logical forms that the present version of the modifier view assigns to these examples as follows:

(49) $[\Delta :: A(t)][B(t)]$.

(50) $\Delta(\exists x([\Box :: A(x)][B(x)]))$.

In the first example, the counterfactual 'if'-clause is taken to modify the probability operator. In the second example, however, the overt probability operator cannot be modified. Rather, the counterfactual 'if'-clause needs to be taken to restrict or modify a different operator. On the semantic level, the second example is then equivalent to assuming the counterfactual to contribute the standard truth conditions (where '$\Box\!\!\rightarrow$' represents, as before, the counterfactual connective of standard semantics):

(51) $\Delta(\exists x(A(x)\Box\!\!\rightarrow B(x))$.

So, we see that on the modifier view, the probability operator does not govern two sentences one of which follows logically from the other. As a consequence, the modifier view cannot provide a rationale for being justified in transiting from the first example to the second. The different roles assumed by the modifier view to be played by the counterfactual in these two constructions blocks the possibility of such a transition. Thus, the modifier view has to assume certain transition failures which are, as far as I can see, simply not there.

Once the general structure of this problem is recognized, it is not hard to find more cases of unwanted transition failures. Consider, for instance, the inference from a counterfactual to a conjunction which conjoins this counterfactual with a trivial conjunct:

(52) It is likely that if the cook had done it, there would have been blood on the kitchen knife.

(53) It is likely that $2 + 2 = 4$ and if the cook had done it, there would have been blood on the kitchen knife.

Assume that the first example is accounted for by taking the 'if'-clause to modify the probability operator. As in the previous case, the second example should not be accounted for by assuming the 'if'-clause to modify the probability operator, for otherwise one would need to move the sentence '$2 + 2 = 4$' into the scope of the 'if'-clause. For this reason, the strategy under present consideration for dealing with embeddings is to assume that the counterfactual conjunct contributes the standard truth conditions. But then we are not justified in transiting from the first to the second example, for we may accept the first example without taking it to be likely that at *all* relevant worlds at which the cook does it, there is blood on the kitchen knife. So, on the modifier view, the application of the probability operator to the conjunction would yield a much stronger probability statement than its application to the single counterfactual. However, the conjunction and the single counterfactual seem to be epistemically equally likely.

Similarly, consider also the following case. On the basis of

(54) It is certain that there is blood on the kitchen knife

and

(55) It is likely that if the cook had done it, there would have been blood on the kitchen knife,

we seem to be entitled to accept

(56) It is likely that there is blood on the kitchen knife and if the cook had done it, there would have been blood on the kitchen knife.[15]

Now, if the likelihood of the counterfactual is explained by assuming 'it is likely' to be modified by the counterfactual 'if'-clause but the counterfactual contributes the standard truth conditions when it is part of the final conjunction, then the conjunction could be unlikely despite the fact that the two starting ascriptions of subjective probability are correct. This consequence is unattractive.

Up to this point, we have discussed the problem for the modifier view as it arises when the modifier view is combined with the standard truth conditions for simple counterfactuals. But a more general conclusion is warranted in the light of the structure of the problem. It seems that the problem is primarily caused by assigning counterfactuals different roles in contexts which may be epistemically closely related. For this reason,

[15] There is an underlying fact about probabilities grounding this entitlement: if $P(A) = 1$ and $P(B) = x$, then $P(A \wedge B) = x$.

we can expect that it will also arise for assignments of truth conditions other than the standard truth conditions as long as these truth conditions are not apt to explain the interaction of counterfactuals with epistemic phrases and so require counterfactual 'if'-clauses to sometimes modify epistemic phrases.

In sum, the situation may be assessed in the following way. The modifier view gains explanatory potential by combining the possibility of explaining the interaction of counterfactuals with epistemic operators as a certain kind of modification while still allowing counterfactuals in other contexts to contribute the standard truth conditions. However, despite its flexibility, this strategy of dealing in two radically different ways with different kinds of embedded counterfactuals has a serious drawback: it cannot explain why we are often justified in transiting between two contexts for which different explanations are given. Even worse, it predicts that we are not justified in doing so.

5.8 Conclusion

Let us take stock. Counterfactuals embed under epistemic operators similar to how indicative conditionals do. Perhaps the currently most popular way of explaining the data in the indicative case is to assume that 'if'-clauses act as restrictors in the problematic cases. This leads naturally to the question whether the interaction between epistemic operators and counterfactuals can be explained in an analogous way. However, the most straightforward explanation fails. Counterfactuals cannot be taken as restrictors in the relevant constructions. But it is possible to extend the restrictor view by allowing counterfactual 'if'-clauses to act as modifiers. As it turns out, a theory of modification can be developed by explaining the operation of domain modification along the lines of imaging. However, the resulting view is not without problems. First, there does not seem to be any independent evidence for the assumption that counterfactual 'if'-clauses can act as modifiers. Secondly, and more importantly, the view still faces a problem of embedding, for it fails to account for natural transitions between contexts in which an 'if'-clause is supposed to act as a modifier of an overt epistemic phrase and an inferentially related context in which it has to be taken to restrict a different covert operator. This problem would not occur if we were able to give a propositional explanation of the way counterfactuals embed under epistemic operators. So, it seems worth exploring possible ways we might explain the problematic data in terms of propositions after all. This will be the task for the upcoming chapter. Nevertheless, the modifier view remains a serious fall-back option should all attempts to give a propositional explanation ultimately fail. In this case, one may simply have to live with its costs, which could perhaps be further reduced by some pragmatic manoeuvres.

6

Counterfactuals and Arbitrariness

Here is where we are. Counterfactuals interact with epistemic phrases in a surprising way reflecting a pattern which suggests that our thinking about counterfactuals is coordinated by a certain conception of conditional counterfactual chances (chapter 3). We can pin this pattern down in the form of an evaluation constraint resembling Lewis's Principal Principle: on the assumption that the counterfactual chances are thus-and-so, our credence in the corresponding counterfactual should be thus-and-so. At this point, problems arise. The evaluation constraint is in apparent conflict with the very possibility of assigning truth conditions to counterfactuals (chapter 4). In the light of this danger, one may start looking for a non-propositional explanation of the relevant data. A theory of this kind emerges from the restrictor view of conditionals (chapter 5). On a generalization of this view, we may assume that counterfactual 'if'-clauses are capable of modifying certain epistemic phrases. On the epistemic level, we may take this to show that the credences in counterfactuals as reported by the evaluation constraint are not ordinary credences in a proposition but rather a special sort of modified credences: credences in the consequent within the scope of the counterfactual supposition of the antecedent. The main problem with this view is that when counterfactuals are embedded into larger contexts, they do not seem to play a different role from when they occur unembedded or within the scope of an epistemic operator. The most natural conclusion to draw from this problem would be to assume that the behaviour of counterfactuals should be given a unified explanation in terms of propositions after all. At this stage, however, we are back where we started, facing the argument that there simply may be no propositions which can do the job.

But we have also seen in our analysis of the puzzle about truth conditions that it rests on an epistemological principle which does not hold in full generality despite being plausible in many cases. If counterfactuals have an epistemic profile which escapes this principle, we may be able to assign them truth conditions after all. In this chapter, we will explore this possible niche. I will present a semantics for counterfactuals on the basis of which I hope to be able to explain the evaluation of counterfactuals. The solution I offer stays fairly close to standard semantics. Instead of the universal quantifier employed in the standard specification of truth conditions, I will suggest using the epsilon-operator which arbitrarily selects a relevant antecedent-world. As we shall see, the epsilon-operator has some special epistemic properties whereby one may escape the arguments against truth conditions for counterfactuals.

6.1 A Proposal

In a search for adequate truth conditions for counterfactuals, the two main options are either truth conditions which are weaker than the standard truth conditions or truth conditions which are at least as strong as the standard truth conditions. Both options are called into question by the general argument against truth conditions (see chapter 4). However, if truth conditions are nonetheless wanted, opting for weaker truth conditions seems to be the more promising possibility, for the argument against weaker truth conditions does not seem quite as watertight as the argument against stronger ones. The argument against weaker truth conditions relies on assumptions about knowability which are incompatible with the existence of unknowable truths. Moreover, they rule out cases in which our epistemic access to a proposition *A* is such that we can come to know *A* only by being in a position to know a proposition *B* stronger than *A*. Applied to counterfactuals, this opens up the possibility that there may be truth conditions for counterfactuals which are weaker than the standard truth conditions but which we can know to obtain only by being in a position to know that some stronger conditions—possibly the standard truth conditions—obtain.

The structure of Stalnaker's (1968) semantics can be seen as providing the basis for truth conditions which are weaker than the standard truth conditions, for if all relevant antecedent-worlds are consequent-worlds, then a particular relevant antecedent-world is a consequent-world (but not *vice versa*). Yet Stalnaker's original background assumption that there is always a single most relevant antecedent-world is implausible.[1] Standard semantics seems to be right in assuming that there are usually many relevant antecedent-worlds. It would be nice to preserve this assumption in an attempt to develop weaker truth conditions along the lines of Stalnaker's semantics.

The basic idea will be to substitute in the standard account the epsilon-operator for the universal quantifier. Intuitively, the epsilon-operator *arbitrarily* selects a world out of the set of relevant antecedent-worlds. A counterfactual will then be true if the consequent is true at the arbitrarily selected antecedent-world. These truth conditions are weaker than the standard truth conditions: if all relevant antecedent-worlds are consequent-worlds, then an arbitrarily selected antecedent-world is a consequent-world, but an arbitrarily selected antecedent-world may be a consequent-world without all antecedent-worlds being consequent-worlds. One can already see how this provides a solution to the puzzle posed by the evaluation of counterfactuals. On the one hand, we can think it likely that an arbitrarily selected antecedent-world is a consequent-world without thinking it likely that all antecedent-worlds are consequent-words. This can happen, for instance, when we think that most but not all antecedent-worlds are consequent-worlds. On the other hand, we can only know that an arbitrarily selected antecedent-world is a consequent-world if we are in a position to know that all antecedent-worlds are consequent-worlds.

The element of arbitrariness present in such a semantics can be motivated by paralleling 'would'-expressions with 'will'-sentences. As remarked in the introductory chapter, it is a common linguistic view that 'would' is derived from 'will' and can be regarded as a specific kind of past tense form of 'will'. For this reason, it is a natural assumption that the semantics of 'would' should be similar to the semantics of 'will'. As 'will' is of intermediate strength between future necessity and future possibility, the modal character of 'would' is likely to be of intermediate strength between counterfactual necessity and counterfactual possibility (see also section 7.2). Now, the semantics for 'will' can be fixed—in some way or

[1] In response to the present problem, Stalnaker (1978) suggests a modification of his theory along supervaluationist lines. We will briefly address this modification in section 6.1.4.

other—by binding it to the one of the possibly many futures which will eventually be actual. A similar approach is not possible in the case of 'would'. With respect to the many relevant counterfactual futures, there is no distinguished one which ultimately outlasts the others, for counterfactual futures do not play themselves out. One way to resolve this conflict is to bind the semantics of 'would' to an arbitrary counterfactual future. Before this line of thought can be developed in more detail, we will need to take a closer look at the epsilon-operator.

6.1.1 *The Epsilon-Operator*

The epsilon-operator was originally introduced by Hilbert as part of his finitistic program.[2] Nowadays, the epsilon-operator is sometimes employed in linguistics to account for *E*-type pronouns and definite noun phrases.[3] Philosophically, the epsilon-operator is of interest, for it is closely related to the use of schematic or arbitrary names in prominent systems of natural deduction. For this reason, issues concerning the epsilon-operator resemble issues in the philosophical debate about arbitrary names and reference to arbitrary objects.

Let us start by saying something about the syntax of the epsilon-operator. The epsilon-operator is a term forming operator on formulas. If $\ulcorner\phi(x)\urcorner$ is a formula, then $\ulcorner\epsilon x\phi(x)\urcorner$ is a term (and not a formula such as $\ulcorner\forall x\phi(x)\urcorner$). As a result, the expression $\ulcorner\phi(\epsilon x\phi(x))\urcorner$ is a well-formed formula. Importantly, the variable x is not free in an epsilon-term like $\ulcorner\epsilon x\phi(x)\urcorner$. The epsilon-operator acts in this respect like a quantifier which binds x similar to a definite description operator.

Semantically, an epsilon-term like $\ulcorner\epsilon x\phi(x)\urcorner$ can be thought of as denoting an arbitrarily selected object satisfying $\ulcorner\phi(x)\urcorner$. But in contrast to definite descriptions, no uniqueness assumption about ϕ is made: an epsilon-term like $\ulcorner\epsilon x\phi(x)\urcorner$ is not taken to be empty if there are many ϕs. To give a semantic model for epsilon-terms like $\ulcorner\epsilon x\phi(x)\urcorner$, we will need a function which selects an object out of the set of ϕs. The epsilon-term can then be taken to denote the object so selected.[4] Formally, a selection function is a function s with the property that $s(A) \in A$ (provided that A

[2] See Hilbert (1922), Hilbert (1923), and also Ackermann (1924). A general discussion of the role of the epsilon-operator in Hilbert's finitistic program can be found in Zach (2003).

[3] See e.g. Slater (1986) and von Heusinger (2004).

[4] This idea stems from Schröter (1956: 46).

is non-empty). Then, if A is the extension of the formula $\ulcorner\phi(x)\urcorner$, we can say that $\ulcorner\epsilon x\phi(x)\urcorner$ denotes $s(A)$.

Based on the present idea, Asser (1957) has developed a comprehensive model theory for the epsilon-operator. To begin with, we take a first-order language and supplement it with the epsilon-operator. We may then define a model $\mathcal{M}$ for this language to be a triple $\langle D, I, s\rangle$ with the following properties:

- D is any non-empty set (the *domain* of discourse).
- I is an *interpretation function* which assigns to any constant an object in D, to any n-ary predicate an n-ary relation over D, and to any n-ary function symbol a function from the n-ary Cartesian product of D into D.
- s is a function $\mathcal{P}(D) \to D$ such that $s(A) \in A$ if $A \neq \emptyset$.

The first two components of such a model are familiar. What is new is the selection function: its sole purpose is to account for epsilon-terms.

The next step would be to define what it means for a formula ϕ to hold in such a model $\mathcal{M}$ relative to a variable assignment β (formally, $\mathcal{M}, \beta \models \phi$). As it turns out, the fact that the epsilon-operator is a term forming operator on formulas complicates the definition considerably. But the basic idea is simple: add to the standard definition a clause which says that epsilon-terms are assigned their semantic value by the selection function. More formally, assume that $\mathcal{M}$ is a model and β a variable assignment. If v is a variable and a an object in the domain, then $\beta\frac{a}{v}$ is a variable assignment which differs from β only in assigning a to the variable v. For a set of sentences Γ and a sentence ϕ, the expression $\ulcorner\Gamma \models \phi\urcorner$ is defined in the standard way. Now, if R_β is the denotation function which assigns to each term a referent, then we will define this function in a way such that the following equation is satisfied:

$$R_\beta(\ulcorner\epsilon v\phi(v)\urcorner) = s(\{a \in D : \mathcal{M}, \beta\frac{a}{v} \models \phi(a)\}).^5$$

Thus, epsilon-terms are assigned referents by applying the selection function s to the set of objects satisfying the formula which makes up the epsilon-term.

[5] Cf. von Heusinger (2004: 313).

Let me briefly comment on epsilon-terms which contain an empty formula. If ϕ is not true of any objects in the domain, then the epsilon-term $\ulcorner \epsilon v\phi \urcorner$ is still assigned a semantic value, namely $s(\emptyset)$. Of course, the selected object cannot satisfy ϕ. To some extent, this may be felt to be counterintuitive: how can a descriptive term be denoting if nothing satisfies the description? However, on the present stipulation, the semantics remains classical which spares us a lot of work.[6] But let me point out that in the application to counterfactuals below, nothing will depend on this convention.

Asser (1957) axiomatized the resulting logic and proved it to be sound and complete. As it turns out, the epsilon-operator is governed by two basic axioms. The first axiom for the epsilon-operator has already been implicitly assumed:

Epsilon. $\exists v\phi(v) \supset \phi(\epsilon v\phi(v))$.[7]

In words: If there is an object satisfying ϕ, then the arbitrary object $\epsilon v\phi(v)$ satisfies ϕ. This is very intuitive: an arbitrary ϕ should always be a ϕ provided that there is a ϕ. This axiom is a theorem according to the present semantics, for the denotation of an epsilon-term is always an element of the extension of the embedded formula as long as this extension is non-empty.

The second basic axiom is this:

Extensionality. $\forall v(\phi(v) \equiv \psi(v)) \supset \epsilon v\phi(v) = \epsilon v\psi(v)$.

Informally: If two formulas are materially equivalent, the corresponding epsilon-terms denote the same object. This ensures that the semantic value of an epsilon-term depends only on the extension of the embedded formula. It is clear that this axiom holds in the present semantics, for the selection function takes only the extension of the relevant formula as an argument.

It is a common thought that a canonical way of proving a universally quantified formula is to show of an arbitrarily selected object that it satisfies this formula. In the light of this, it is a natural question whether a

[6] Asser (1957) also contains a treatment of epsilon-terms with the aid of partial selection functions which are undefined for the empty set.

[7] Assume that $\ulcorner \exists v\phi(v) \urcorner$ is a closed formula (a sentence). If we allow it to be open, a proviso is necessary to avoid an unwanted interplay of the variables in the epsilon-term and the quantifiers in the original formula. Cf. Asser (1957: 42).

universally quantified sentence $\ulcorner \forall x \phi(x) \urcorner$ is implied by $\ulcorner \phi(\epsilon x(x = x)) \urcorner$. The question is whether 'Everything is a ϕ' is implied by 'An arbitrary object is a ϕ'. When we formalize this latter sentence, we somehow need to express that the selection of objects is unrestricted. A good way of doing so seems to be to take a tautologous formula such as $\ulcorner x = x \urcorner$ which is logically guaranteed to be true of everything.

It is not hard to see, though, that this implication does not hold on the present semantics. This is because all that is required for an epsilon-formula such as $\ulcorner \phi(\epsilon x(x = x)) \urcorner$ to be true in a model is that the selection function picks an object which is ϕ. But this is possible without all objects being ϕs. To put the point slightly differently: an epsilon-formula such as $\ulcorner \phi(\epsilon x(x = x)) \urcorner$ is construed as a formula about a particular object, namely $\epsilon x(x = x)$, and this object may be ϕ without all objects being ϕ.

Nevertheless, there is a more qualified relation between epsilon-terms and the universal quantifier which can do justice to the convention of proving a universally quantified formula by proving—given certain side conditions—of an arbitrary object that it satisfies the formula:

> **Epsilon-Terms and Generality.** Let Γ be a set of sentences which do not contain the epsilon-operator and let $\phi(x)$ be a formula which does not contain the epsilon-operator either. Then
>
> $$\Gamma \models \phi(\epsilon x(x = x)) \text{ iff } \Gamma \models \forall x \phi(x).$$

Less formally: If it follows from a set of sentences that an arbitrary object is ϕ, then it also follows that everything is ϕ and *vice versa*. The side conditions should be no surprise: they exclude that any particular assumptions about $\epsilon x \phi(x)$ enter the stage.

A remark on how the present observation can be proven. Take a model $\mathcal{M}$ which verifies Γ. Since Γ does not contain the epsilon-operator, any model which differs from $\mathcal{M}$ only in the choice of the selection function will verify Γ as well. So, $\mathcal{M}$ verifies $\ulcorner \phi(\epsilon x(x = x)) \urcorner$ no matter which object is assigned to the epsilon-term $\ulcorner \epsilon x(x = x) \urcorner$. Since ϕ does not contain the epsilon-operator either, the extension of ϕ is independent of the semantic values of epsilon-terms. Hence, $\mathcal{M}$ verifies $\phi(x)$ no matter which object is assigned to its free variable showing $\ulcorner \mathcal{M} \models \forall x \phi(x) \urcorner$. So, if $\Gamma \models \phi(\epsilon x(x = x))$, then $\Gamma \models \forall x \phi(x)$. The other direction is an immediate consequence of universal instantiation.[8]

[8] One may also wonder whether the same kind of equivalence holds between $\ulcorner \forall x(\phi(x) \supset \psi(x)) \urcorner$ ('All ϕ-s are ψ-s') and $\ulcorner \psi(\epsilon x \phi(x)) \urcorner$ (An arbitrary ϕ is a ψ'). But when the universal

The relation between epsilon-terms and generality is reflected in the epistemic profile of sentences containing epsilon-terms. We can be justified to a high degree that an arbitrarily selected object has a certain property without being equally justified in thinking that everything has this property. If we know that most but not all objects have a certain property, we are justified to a high degree that an arbitrarily selected object has this property. However, we are not justified in thinking that all objects have this property. The possibility of having high credence in $\ulcorner\phi(\epsilon x(x = x))\urcorner$ without having high credence in $\ulcorner\forall x\phi(x)\urcorner$ reflects the fact that a sentence of the form $\ulcorner\phi(\epsilon x(x = x))\urcorner$ does not imply $\ulcorner\forall x\phi(x)\urcorner$. The situation changes when we consider knowledge. If we are in a position to know that an arbitrarily selected object has a certain property, we are in a position to know that everything has this property (provided we do not know anything particular about the object in question). Logically, this reflects the fact that if we can prove $\ulcorner\phi(\epsilon x(x = x))\urcorner$ (without making any specific assumptions about the epsilon-term), we can prove $\ulcorner\forall x\phi(x)\urcorner$.

The epistemic profile of sentences containing epsilon-terms is exactly of the kind we are looking for. There are logically stronger sentences, namely the corresponding universally quantified sentences, such that (a) we can be justified in assigning a high subjective probability to the epsilon-sentences without assigning a high subjective probability to these stronger sentences and (b) we can only know the epsilon-sentences to be true if we are in a position to know the corresponding stronger sentences to be true. Thus, epsilon-terms stand to the corresponding universally quantified sentences in the same epistemic relation as counterfactuals seem to stand to the standard truth conditions which are specified by a universal quantifier. For this reason, it seems promising to see whether we can give a semantics of counterfactuals by using the epsilon-operator.

6.1.2 Arbitrary Truth Conditions

Equipped with the epsilon-operator, we are able to talk about arbitrary worlds in our semantics. For instance, given a world w and a set of worlds A, $\epsilon w'(w' \in f_w(A))$ is an arbitrary relevant A-world (relative to w). With this possibility in mind, the proposal I would like to make can be stated as follows:

quantification is vacuously true, the epsilon-sentence need not hold. However, once Γ proves $\ulcorner\exists x\phi(x)\urcorner$, the equivalence holds. This problem is avoided in the present theorem, for in classical logic, $\ulcorner\exists x(x = x)\urcorner$ is a logical theorem and therefore implied by any set of assumptions.

Epsilon-Based Truth Conditions. Let w be any world and $\ulcorner \phi > \psi \urcorner$ a counterfactual. Then $w \models \ulcorner \phi > \psi \urcorner$ iff

$$(\exists w' \in f_w(\llbracket \phi \rrbracket)) \supset (\epsilon\, w'(w' \in f_w(\llbracket \phi \rrbracket)) \models \psi).$$

In words: A counterfactual is true at a world w if either there is no relevant antecedent world or, if there is one, the consequent is true at the arbitrarily selected relevant antecedent-world (relative to w).

A few remarks on how these truth conditions are fixed. They are specified by a material conditional, the antecedent of which expresses that there is a relevant antecedent-world. If there is none, the counterfactual comes out true. Thus, the present truth conditions share with the truth conditions of standard semantics the feature that counterfactuals with an impossible antecedent are vacuously true. This feature is to a large extent a side effect of the present account, for it is brought about by a clause specifically designed for this case. *Prima facie*, there is no reason why the core of the present semantics could not be extended in a different way to allow some counterfactuals with an impossible antecedent to be false.[9]

The main part of the specification of truth conditions is contained in the consequent of the material conditional. The selection function f takes the world w and the set $\llbracket \phi \rrbracket$ of worlds at which the antecedent is true as arguments and provides the set $f_w(\llbracket \phi \rrbracket)$ of relevant antecedent-worlds. Now, the epsilon-operator arbitrarily selects one of these worlds. The whole clause is then true if the consequent is true at this arbitrarily selected world.

To compare the present semantics to the standard semantics, let us recall the formal specification of the standard truth conditions:

Standard Semantics. Let w be any world and $\ulcorner \phi > \psi \urcorner$ a counterfactual. Then $w \models \ulcorner \phi > \psi \urcorner$ iff

$$\forall w'(w' \in f_w(\llbracket \phi \rrbracket) \supset w' \models \psi).$$

The present kind of truth conditions is indeed weaker than the standard truth conditions. Suppose that the standard truth conditions are satisfied. If this is because there are no relevant antecedent-worlds, then the

[9] For a recent and extensive defense of the claim that counterpossibles are always true, see Williamson (forthcoming). Considerations against can be found in Nolan (1997) and Brogaard and Salerno (2013). See also Fine (2012a,b) for potential counterexamples against the related claim that counterfactuals do not create hyper-intensional contexts.

arbitrary truth conditions are satisfied as well. Otherwise the set $f_w([\![\phi]\!])$ will be non-empty and all worlds in it will verify the consequent of the counterfactual. But then an arbitrarily selected world out of this set has to verify the consequent. So, the standard truth conditions imply the epsilon-based truth conditions. The converse will generally not be the case, for an arbitrarily selected antecedent-world may verify the consequent without all antecedent-worlds doing so.

Structurally, we have arrived at a kind of Stalnaker semantics. Whether or not a counterfactual is true generally depends on whether the consequent is true at a single world. Stalnaker's original theory is better described with a definite description operator: the truth conditions of counterfactuals are given by what is true at *the most relevant antecedent-world*. As already noted, the problem for Stalnaker's original suggestion is that there is only rarely, if ever, a single most relevant antecedent-world. On the present account, this difficulty is avoided. It is actually an integral part of the present theory that there are usually many relevant antecedent-worlds. Stalnaker (1978) refined his original account by supervaluating over all the candidates for being the closest antecedent-world. From the perspective of the present theory, the precisifications picking a unique closest antecedent-world correspond to the possible arbitrary selections of an antecedent-world out of the set of relevant antecedent-worlds (see also section 6.1.4).

6.1.3 The Logic of Counterfactuals

So far, I have given an outline of an alternative semantics for counterfactuals. What remains to be done is to show how the logic of counterfactuals can be approached within the present framework. In standard semantics, the logic of counterfactuals can be adjusted by imposing constraints on the selection function f which provides the set of relevant antecedent-worlds (we have looked at these constraints in section 3.1). Since the present semantics is given partly in terms of this function, we can influence the logic in this way as well by constraining the sets from which the epsilon-operator selects its referents. As we shall see, one may additionally want to consider introducing a certain kind of coordination constraint which is directly concerned with the epsilon-operator.

Let us start by seeing how the constraints on the selection function influence the logic of counterfactuals. One requirement is that relevant antecedent-worlds are always antecedent-worlds, i.e. $f_w(A) \subseteq A$.

This secures the logical truth of ⌜$A > A$⌝, for if relevant antecedent-worlds are always antecedent-worlds, then an arbitrarily selected relevant antecedent-world will be an antecedent-world. If $A = \emptyset$, i.e. the antecedent is impossible, the constraint implies that $f_w(A) = \emptyset$ as well, i.e. that there are no relevant antecedent-worlds. This makes counterfactuals with an impossible antecedent vacuously true.

Another constraint requires that the selection function selects some antecedent-worlds unless there are no antecedent-worlds, i.e. $f_w(A) = \emptyset$ implies $A = \emptyset$. Just like in standard semantics, this constraint is important for our specification of truth conditions, too, as it guarantees that only counterfactuals with an impossible antecedent are vacuously true.

Next there are the centering conditions. Recall that weak centering requires that if an antecedent A is true at a world w, then $w \in f_w(A)$, which implies that if the antecedent is true at the actual world, then the actual world is a relevant world. Strong centering makes in this case the actual world the only relevant world, or, more generally, it requires $\{w\} = f_w(A)$ if $w \in A$. Perhaps the strongest reason for imposing the centering conditions is to ensure the validity of *modus ponens* (we discuss another issue about centering in section 7.1). On standard semantics, weak centering is sufficient for this purpose because if A is true at the actual world and if B is true at all relevant A-worlds, then B is true at the actual world, too, given that the actual world is a relevant A-world. On the present semantics, we would have to adopt strong centering, however. In this case, if the antecedent is true at the actual world, the selected world will be the actual world as it is the only possible choice and so the truth of the corresponding counterfactual will require the truth of its consequent. Weak centering, on the other hand, would leave open the possibility that the selected antecedent-world is not the actual world and so modus ponens could fail. If one favours weak centering over strong centering as a description of the structure of relevance, one would need to impose an additional coordination constraint on the epsilon-operator (see below) or else accept failures of *modus ponens* for counterfactuals (which strikes me as the less attractive option).

If one compares the logic reached so far to Stalnaker's (1968) logic C2 or to Lewis's (1973) official logic VC (which is weaker than the present logic in other respects), then it turns out that there is precisely one interesting principle missing (and some consequences of it). This principle states that counterfactually equivalent propositions can be substituted for each other in the antecedent of a counterfactual:

Substitution. $A > B, B > A, A > C \models B > C.$

This principle is quite complex and cannot easily be tested against intuitions. After an early attempt by Tichy (1978) to give a counterexample and a response by Stalnaker (1984: Ch. 7), this principle and its semantic justification have recently been put to closer scrutiny. Tooley (2002) discovered a certain type of counterexample to it the consequences of which have been further explored by Cross (2006).[10] The counterexample by Tooley requires, however, the possibility of backward causation. If one takes the counterexample to be genuine, one might simply stop here and leave the logic as it is (this is the recommendation in Cross 2009).

Although I find the counterexample by Tooley *prima facie* very convincing, there are two considerations which speak in favour of having a backup plan. The first one is simply that it would be nice to prevent a semantics of counterfactuals from being hostage to the fortune that backward causation is really possible. A related thought starts with the observation that there will in any case be large parts of metaphysical space which are free from backward causation. Unless the counterexample can be generalized to cases not involving backward causation, we would expect the substitution principle to hold in such a restricted metaphysical environment. A good theory of counterfactuals should have the resources to bolster this expectation.

In standard semantics, the substitution principle is validated because the selection function satisfies a constraint which was already mentioned in section 3.1. To recall:

Equivalence for Relevance. Let w be any world, A and B be sets of worlds. If $f_w(A) \subseteq B$ and $f_w(B) \subseteq A$, then

$$f_w(A) = f_w(B).$$

Now, if the relevant A-worlds are the relevant B-worlds, then the truth of $\ulcorner A > C \urcorner$ implies the truth of $\ulcorner B > C \urcorner$ and so the substitution principle turns out to hold.

[10] Ahmed (2011) also proposes a counterexample. But see Walters (2011) for a case against it. Further discussion which is particularly relevant for the epsilon-based semantics can be found in Bacon (ms).

Actually, this equivalence condition is the consequence of a more general principle holding in standard semantics:

Generalized Equivalence. Let w be any world, A and B be sets of worlds. If $f_w(A) \cap B \neq \emptyset$ and $f_w(B) \cap A \neq \emptyset$, then

$$f_w(A) \cap f_w(B) = (f_w(A) \cap B) \cup (f_w(B) \cap A).$$

If the starting conditions of the original equivalence constraint are satisfied, then $f_w(A) \cap B = f_w(A)$ and $f_w(B) \cap A = f_w(B)$. Hence, the generalized principle would give us $f_w(A) \subseteq f_w(B)$ and $f_w(B) \subseteq f_w(A)$ showing that the original principle is indeed implied by the generalized one.[11]

To see why the generalized equivalence principle is a consequence of the standard account, recall that the closeness relation is assumed to be total in the sense that any two worlds are comparable with respect to their closeness to a given third world (they may, of course, still be equally close). The function f then takes a world w and a proposition A to the set of closest A-worlds with respect to w. Now, if some of the closest A-worlds are B-worlds and some of the closest B-worlds are A-worlds, these $(A\&B)$-worlds are all comparable to each other. If one of those worlds, w', were part of $f_w(A)$ but not of $f_w(B)$, say, then w' would either be closer to w than the $f_w(B)$-worlds or further away from them. In the first case, $f_w(B)$ would not consist of the closest B-worlds, for w is a B-world. In the second case, $f_w(A)$ would not consist of the closest A-worlds, for $f_w(B)$ contains A-worlds which are closer than w'. Thus, we see that the substitution principle effectively derives from the idea that the closeness relation is total combined with the general assumption that the truth of a counterfactual depends on what is true at the closest antecedent-worlds (cf. Cross 2006, 2008).

Let us now turn to how the substitution principle fares on the present semantics. To clarify its relation to the equivalence condition a little further, suppose that the counterfactuals $\ulcorner A > B \urcorner$ and $\ulcorner B > A \urcorner$ are true. This would mean that the arbitrarily selected relevant A-world is a B-world $(\epsilon w'(w' \in f_w(A)) \in B)$ and the arbitrarily selected relevant B-world is an A-world $(\epsilon w'(w' \in f_w(B)) \in A)$. Now, if we further assume

[11] Hans Rott pointed out to me that the somewhat clumsy equivalence condition is provably equivalent to the conjunction of two more natural conditions known from rational choice theory and usually referred to as α and $\beta+$. I regret that I have not yet been able to explore these parallels further.

that the counterfactual $\ulcorner A > C \urcorner$ is true, we get the consequence that $\ulcorner B > C \urcorner$ is true just in case our assumptions ensure the identity of the selected A-world with the selected B-world, for only then does the selected A-world being a C-world ensure that the selected B-world is a C-world as well. Now, if the equivalence constraint is in place, we get the required identity implication in the special case in which all relevant A-worlds are B-worlds and all relevant B-worlds are A-worlds. Then the relevant A-worlds will coincide with the relevant B-worlds by the equivalence constraint. Since these sets are identical, the selected relevant A-world will be the selected relevant B-world by the extensionality of epsilon-terms. However, if some of the relevant A-worlds are B-worlds but some are not, then the selected relevant A-world might be a B-world without all relevant A-worlds being B-worlds. In such a situation, the equivalence constraint will not be applicable and the relevant A-worlds will be distinct from the relevant B-worlds as some relevant A-worlds are not B-worlds. Still, some relevant B-worlds may be A-worlds and so the selected B-world may be an A-world without the selected worlds being identical. But then the substitution principle could fail.

The counterexamples to the substitution principle which would still occur even if the equivalence constraint on f is in place are somewhat peculiar. They can come about only in situations in which a counterfactual $\ulcorner A > B \urcorner$ is true without all relevant A-worlds being B-worlds. As explained below, the latter condition will prevent the original counterfactual $\ulcorner A > B \urcorner$ from being known. Given that the additional counterexamples to the substitution principle would occur only in such specific circumstances, they may be tolerable despite lacking—as far as I can presently see—intuitive support (they cannot be justified in a way similar to Tooley's counterexample).

The present defence can be generalized to show that a possible damage caused by the present semantics will always be confined to a limited range of cases. Call a counterfactual *determinately true* iff all relevant antecedent-worlds are consequent-worlds (see also the discussion in section 6.1.4). Regarding determinate counterfactuals, the epsilon-operator has no choice except selecting a world at which the consequent is true. Now, given that determinate truth is truth according to standard semantics and the logic in standard semantics can be fixed by constraints on the selection function f, we can, on the present account, recover a possible logic holding for some version of standard semantics as a possible logic for determinate truth. This is because the present semantics is partly given in terms of the same kind of selection function f and so we can adopt whichever constraints seem reasonable from the perspective of standard semantics. Determinate truth will then obey whichever logic seems plausible from the perspective of standard semantics. Hence, counterexamples to some logical principles which may not hold on the present semantics but seem plausible from the perspective of standard semantics can, by this strategy, be ensured to occur only in cases in which truth is not determinate.

If one wishes to avoid these additional counterexamples to the substitution principle, one would need to introduce a new kind of *coordination constraint* directly concerned with epsilon-terms. One way of doing this is to constrain the choices open to the epsilon-operator in conditions pertaining to the validity of the substitution principle. In order to deal with possible counterexamples to the substitution principle, such a constraint should not be too strong and would ideally go hand in hand with the equivalence constraint for the relevance function f. Here is a suggestion:

Modest Equivalence for Choice. Let w be any world, A and B be sets of worlds. If $\epsilon w'(w' \in f_w(A)) \in f_w(B)$ and $\epsilon w'(w' \in f_w(B)) \in f_w(A)$, then

$$\epsilon w'(w' \in f_w(A)) = \epsilon w'(w' \in f_w(B)).$$

In words: If the arbitrary relevant A-world is a relevant B-world and the arbitrary relevant B-world is a relevant A-world, then the selected worlds are identical. If we combine this constraint with the generalized equivalence condition, we can show the substitution principle to hold. Suppose $\ulcorner A > B \urcorner$ and $\ulcorner B > A \urcorner$ are true. Given our semantics, this requires that the arbitrarily selected relevant A-world is a B-world ($\epsilon w'(w' \in f_w(A)) \in B$) and the arbitrarily selected relevant B-world is an A-world ($\epsilon w'(w' \in f_w(B)) \in A$). By the generalized equivalence constraint, we find that the selected A-world is actually a relevant B-world and the selected B-world a relevant A-world by being in $f_w(A) \cap f_w(B)$. The modest equivalence condition on choice now gives us the identity of the selected worlds and so adding the premiss $\ulcorner A > C \urcorner$ would suffice for the truth of $\ulcorner B > C \urcorner$.

Adopting only something like the modest equivalence constraint on choice and feeding it in the generalized equivalence condition on relevance allows for a satisfying treatment of the substitution principle. If the substitution principle holds universally, the two constraints secure its validity. If, however, it has counterexamples like the one suggested by Tooley, this can be assumed to be the case because the equivalence condition on relevance fails: sometimes the relevant *A*-worlds are *B*-worlds and the relevant *B*-worlds are *A*-worlds, but the two sets of relevant worlds are none the less disjoint. In this situation, the antecedent for the modest equivalence condition on choice cannot be satisfied and so the substitution principle will fail. Accordingly, the likely truth of the substitution principle in most but possibly not all situations could be accounted for by the truth of the equivalence condition on relevance in most but not all situations.

Adding a coordination constraint like the present one to the picture may constitute a certain cost, for it subtracts somewhat from the idea of having a choice operator which selects its referents on a random basis. Once a coordination constraint is in place, certain choices will no longer be independent of each other. For example, selecting a relevant *A*-world which happens to be a relevant *B*-world may require selecting this world also as the relevant *B*-world under certain circumstances. Clearly, the choices would not be determined, but the whole process would perhaps be better seen as the arbitrary selection of a whole structure representing an assignment of admissible choices rather than a number of (independent) individual choices. If one does not like the introduction of a coordination constraint, one could trade it in for a certain kind of additional counterexample to the substitution principle as described above. For the time being, I would like to leave open which line would be best to take.

Incidentally, if one adopts a coordination constraint in order to deal with the substitution principle, one could secure the validity of modus ponens merely on the basis of weak centering by imposing another coordination constraint. One would require the epsilon-operator to select the world of evaluation if the antecedent is true at it. Weak centering would ensure that the world of evaluation is in such cases part of the relevant worlds. A coordination constraint could then do the work of strong centering by forcing the epsilon-operator to select the world of evaluation out of the set of relevant worlds. This strategy would prevent counterfactual chances from trivializing for true antecedents. If strong centering is adopted, then the counterfactual chances under a true antecedent are always 1 or 0, for they will result from conditionalizing on a singleton set consisting of the actual world only. Now, I do not consider this to be much of a problem, for counterfactual chances are derived from how we evaluate counterfactuals. For this reason, I think that the seemingly unnatural consequence of them being trivial for true antecedents does not show anything which could be assessed on independent grounds. The situation would be different if we were concerned with objective physical chances, for it is clear that they can be non-trivial even when conditionalized on true antecedents—the objective chance of my team losing the match given that they miss the penalty may be high but not 1 even if, as a matter of fact, they are going to miss the penalty. But given the difference between the two conceptions of chance, I think adopting strong centering is not necessarily a problem, but if desired, one could avoid it with the help of a coordination constraint.

6.1.4 The Metaphysics of Arbitrariness

There remains an important issue which I have sidestepped so far: what metaphysical assumptions underlie the use of epsilon-terms in the present semantics? Answering this question has important implications for the metaphysics of counterfactuals. I will only give a brief sketch of some of the available options. To address the issue properly, we would need to venture into the debate about arbitrary reference. In principle, any account of arbitrary reference may be adopted and applied to the present semantics.

According to our semantics, there is a selection function which selects for every non-empty set of worlds one of its members. An epsilon-term $\ulcorner \epsilon x \phi(x) \urcorner$ is taken to denote the object which the selection function selects from the set of objects satisfying $\ulcorner \phi(x) \urcorner$. This is enough to explain how epsilon-terms function semantically and, in particular, in which logical relations they stand to other sentences. However, adopting this semantics presupposes that there is a selection function which provides epsilon-terms with referents. What is the metaphysical status of this assumption?

According to a *realist stance*, there simply is such a selection function which supplies epsilon-terms with referents.[12] In its application to counterfactuals, the realist stance corresponds to the assumption that there are

[12] A realist theory of arbitrary reference is defended in Breckenridge and Magidor (2012). An earlier presentation of this paper at the Ockham Society in Oxford 2007 was an important source of inspiration for the present account.

brute counterfactual facts. An *anti-realist* may object that there cannot be a distinguished selection function, for one selection function is as good as any other. For this reason, epsilon-terms could not be assumed to have determinate referents reflecting brute counterfactual facts. Perhaps it is merely *as if* they denote a certain object in an arbitrary way. Other anti-realist stories may be possible as well.

On a realist conception, epsilon-terms are taken at face value: they are singular terms denoting a particular object, the fact that the descriptive material of an epsilon-term underdetermines this object notwithstanding. Epsilon-terms would pick out their referents in a truly arbitrary way. Of course, the way they pick out their referents would not be completely arbitrary, for the descriptive material partially constrains which objects they might select (an arbitrary F is always an F).

The main challenge for the realist is the question of how epsilon-terms are supposed to get their referents. Since epsilon-terms are construed as picking out their referents in an arbitrary way, a realist will not want to specify an additional condition which—together with the predicative component—*determines* a referent. To some extent, a realist will have to assume that the semantic gods sometimes gamble: the linguistic meaning of an epsilon-term alone does not determine a referent and so the assignment of a referent is to some extent an arbitrary affair.

To guard against a possible misunderstanding, note that the realist conception of epsilon-terms does not imply that epsilon-terms have different referents on different occasions of use. The choice of the selection function which is the semantic value of the epsilon-operator is made once and for all. If one wishes, one could relax this assumption, but in the context of counterfactuals, this only seems to create problems, for then two utterances of the same counterfactual could be such that one is true while the other is false even though the counterfactual does not contain any context-sensitive elements. In principle, there is also room for a middle position according to which the referent of an epsilon-term is only held fixed over the course of a particular conversation.[13] But for present purposes, I assume that the referents of epsilon-terms stay constant.

On the realist conception, sentences containing the epsilon-operator are generally either true or false. In its application to counterfactuals,

[13] The latter option might square well with an implementation of the present semantics into discourse representation theory.

we will find, however, that the truth of certain counterfactuals is sometimes not grounded in particular matters of fact together with the laws of nature. Particular matters of fact and the laws of nature will in general only determine the set of relevant antecedent-worlds. However, if not all relevant antecedent-worlds are consequent-worlds but the arbitrarily selected world is a consequent-world, then the counterfactual is true not only because of what has happened at the actual world and what its laws are but also because of the arbitrary choice of the antecedent-world. So, there will be some arbitrary counterfactual facts which do not supervene on particular matters of facts together with the laws of nature.[14]

How high this metaphysical cost is taken to be depends on one's general metaphysical outlook. If one subscribes to anything like Humean supervenience, realism about epsilon-terms is untenable. But if one allows the ultimate supervenience base to be fairly large, one may be more inclined to include counterfactual facts in them as well. For instance, one may already hold that objective chances do not supervene on physical facts: it may seem possible that two worlds unfold in just the same way but are in fact governed by different objective chance distributions. One world may unfold in a way which is quite probable from the perspective of its objective chance distribution, while the other world may unfold in a way which is improbable in the light of its objective chance distribution. If basic facts concerning objective chances are granted, basic counterfactual facts might seem less implausible.

As an alternative to the realist interpretation, one might endorse an *anti-realist* conception of epsilon-terms. If the predicative component of an epsilon-term is satisfied by multiple objects, the epsilon-term would not be taken to denote one of these objects. Still, epsilon-terms could be taken to purport to denote an object which has been selected in such a way, but they would not be construed as being capable of really doing so. On one way of understanding such an anti-realist conception, epsilon-terms would be used *as if* the realist construal of them were correct.

[14] There are some suggestions in the literature which go some way towards the realist interpretation of the present theory. At the end of his paper, Hawthorne (2005) makes a brief conjecture that the problems surrounding counterfactuals and chance are perhaps best addressed by postulating that there is always a most relevant antecedent-world. McDermott (1999) offers a somewhat related proposal: although he does not defend a Stalnaker semantics, he takes certain counterfactuals to introduce a realm of unknowable facts.

The main challenge for an anti-realist conception of epsilon-terms is to give an indication of how epsilon-terms function semantically if they do not denote. I anticipate that there will be more than one option, but here I will sketch only one possibility which employs the familiar technique of supervaluations.[15] Application of it will relate the present semantics of counterfactuals to Stalnaker's supervaluationist theory of conditionals.

Originally, the technique of supervaluations was developed by van Fraassen (1966) as a tool to treat empty names. Hence, it is a natural question whether epsilon-terms can be accounted for by supervaluationist methods, for an anti-realist takes epsilon-terms, strictly speaking, to be generally empty, the only exception being the case in which the extension of the predicative component contains exactly a single object. On a supervaluationist account, sentences are taken to be either *determinately true*, *determinately false*, or *indeterminate*. It is not hard to see how a natural application of a supervaluationist approach to epsilon-terms might look. The idea would be that a sentence containing an epsilon-term is determinately true iff it is true no matter which object we select out of the extension of the predicative component of the epsilon-term. Likewise, it is determinately false iff we cannot find an object in the extension of the predicative component which would make the sentence true. Otherwise, the sentence would be indeterminate in truth-value.

If we combine the supervaluationist account of epsilon-terms with the proposal for counterfactuals in terms of epsilon-terms, we will get the following picture. Counterfactuals will either be determinately true, determinately false, or indeterminate. They will be determinately true iff all relevant antecedent-worlds are consequent-worlds. So, the standard truth conditions would turn out to be the conditions under which a counterfactual is determinately true. However, a counterfactual will only be determinately false if no relevant antecedent-world is a consequent-world. If some relevant antecedent-worlds are consequent-worlds and some are not, the counterfactual will be said to be indeterminate. In a typical situation, a counterfactual like 'If the coin had been tossed, it would have landed heads' would be indeterminate. Such an account seems to be a little more complex by involving epsilon-terms but otherwise it is

[15] Alternatively, one could give a fictionalist story about epsilon-terms according to which they are used by pretending that they select objects on a random basis.

very similar to Stalnaker's (1978) supervaluationist theory of conditionals (cf. also Lewis 1973: 80).

What happens to the explanation of our credences in counterfactuals if a supervaluationist account of epsilon-terms is adopted?[16] The explanation that has been given is based on the idea that evaluating counterfactuals can be understood as thinking about arbitrary antecedent-worlds. To me, it is at least not clear whether this explanation presupposes the realist construal of epsilon-terms, for the question of what to think about an arbitrarily chosen antecedent-world seems to be well posed independently of how we account for epsilon-terms. The worry could be that if epsilon-terms turn out to be non-denoting and are accounted for on supervaluationist lines, one will need to say something about how credences interact with possibly indeterminate sentences. One option is to say that our credences estimate whether an arbitrary supervaluation verifies the sentence in question. In a similar spirit, one may hold that credences in sentences which are given a supervaluationist semantics estimate the fraction of supervaluations which verify the relevant sentence among all admissible supervaluations, possibly weighted by a certain probability distribution.[17] Applied to counterfactuals, we would then find that our credences estimate (a) whether an arbitrary supervaluation picking out a relevant antecedent-world is a consequent world or (b) what the fraction is of supervaluations verifying the counterfactual among all supervaluations. In both cases, the right kind of credence seems to be predicted.

Supervaluationism allows us to underpin an anti-realist understanding of epsilon-terms. Such a story is not without costs: we would have to live with the fact that the semantics of counterfactuals is given within a meta-language (the language containing the epsilon-operator) for which we can offer only a non-classical (supervaluational) meta-semantics. Thus, the resulting theory of counterfactuals must ultimately be said to be non-classical: some counterfactual would neither be determinately true nor determinately false. For this reason, I would prefer the realist story if it does not turn out to be untenable (and I will continue to write as if the realist story were correct).

16 Thanks to Tobias Rosefeldt for pressing me on this.

17 For further discussion, see Williams (2009: 163–7).

6.1.5 *Comparisons*

Structurally, and independently of whether we adopt a realist or an anti-realist attitude towards epsilon-terms, the present account is a kind of Stalnaker semantics. Whether or not a counterfactual is true depends on whether the consequent is true at a single world. As mentioned before, one difference is that Stalnaker's original theory is more naturally described with a definite description operator: the truth conditions of counterfactuals are given by what is true at *the most relevant antecedent-world*. In contrast, it is an integral part of the present theory that there are usually many relevant antecedent-worlds for the epsilon-operator to select. But we could reinterpret Stalnaker's theory by identifying the most relevant antecedent-world with the arbitrarily selected world. Would such a theory be different from the present model? As far as the modal profile of the proposition expressed by counterfactuals goes, the answer is quite clearly 'No': both theories suggest that a counterfactual $\ulcorner A > B \urcorner$ is true at a set of worlds of the form $\{w : s_w(A) \models B\}$, where the function s selects the relevant antecedent-world. However, the present theory contains a further component which equips counterfactuals with a specific epistemic profile provided by the specification of their truth conditions in terms of the epsilon-operator. As we shall see, the explanation of credences in counterfactuals relies on the relevant antecedent-world being arbitrarily selected from a larger set of relevant antecedent-worlds.

The general idea of accounting for the connection between conditionals and conditional probabilities in terms of arbitrary selection has at least one prominent predecessor. Van Fraassen (1976) (see also Bacon 2015) obtained a limited tenability result for the thesis that probabilities of conditionals are conditional probabilities by considering possible worlds models in which the order of worlds is such that the closest antecedent-worlds are positioned as if they were arbitrarily selected from the set of relevant antecedent-worlds. This result is achieved by extending a standard possible worlds model so that every world in the new model is an infinite series of worlds in the original model and then adjusting the order of these worlds in such a way that the closest antecedent-world can be seen as representing an arbitrarily selected antecedent-world.[18] Less similar,

[18] As van Fraassen was concerned with a different thesis applying to indicative conditionals, we would still need to see whether his result can be adapted to the evaluation of counterfactuals.

but still close in spirit is the random variable approach pursued by Jeffrey (1991) and further explored by Jeffrey and Stalnaker (1994) (see also Kaufmann 2005). The observation is that a conditional probability can be seen as the expected value of a random variable which assumes intermediate values matching the relevant conditional probability at worlds at which the antecedent is false.[19] Given that credences in classical propositions can be seen as the expected value of a random variable which only takes the values 1 and 0 depending on whether the proposition is true or false at the world in question, the random variable approach can be seen as an extension of the classical framework of propositions. However, conditionals would no longer be bivalent. They could have all values in the interval $[0, 1]$. An advantage of the present account is that it is more conservative by allowing conditionals to behave like ordinary propositions.

6.2 The Evaluation of Counterfactuals

A major task for the present semantics is to explain the puzzling credences in counterfactuals reflected in the evaluation constraint developed in chapter 3. The core idea is to explain our credences in counterfactuals by the specific epistemic profile of epsilon-terms: the more antecedent-worlds are consequent-worlds, the more likely should we take it to be true that an arbitrary antecedent-world is a consequent-world. If successful, this would show that the evaluation constraint can, despite appearances to the contrary, be explained on the basis of certain kinds of truth conditions for counterfactuals. In the discussion to follow, I will set two special cases aside: counterfactuals with a true antecedent (see section 7.1) and counterfactuals with an impossible antecedent.

Sentences containing epsilon-terms have a specific epistemic profile. Consider sentences of the form $\ulcorner G(\epsilon x F(x)) \urcorner$ saying that an arbitrary F is G. What are their subjective probabilities? It seems that our subjective probabilities concerning such sentences should be sensitive to what we take to be the distribution of Gs among the Fs, provided we assume that

[19] The random variable approach targets, like van Fraassen, only Adams's thesis. But it seems to have a natural extension to counterfactuals: where the evaluation of indicative conditionals might consist in estimating a random variable corresponding to our present subjective probabilities, counterfactuals might be evaluated by estimating a random variable corresponding to counterfactual chances.

there is at least one F. The more Fs we take to be Gs, the more likely should we take an arbitrary F to be G. If we take few Fs to be G, we should think it quite unlikely that an arbitrary F is G. Similarly, if we take most Fs to be Gs, we should think it quite likely that an arbitrary F is G. In sum, how likely we should take an arbitrary F to be G seems to be coordinated by an estimate of the fraction of Gs among the Fs. Thus, epsilon-terms induce a certain structure on the subjective probabilities we should assign to sentences containing them: our subjective probabilities should be distributed uniformly over the different candidates which the epsilon-operator might select.

The relevant sentence in the specification of arbitrary truth conditions for counterfactuals is of the form $\ulcorner G(\epsilon x F(x)) \urcorner$. It reads: $\epsilon w'(w' \in f_w(\llbracket \phi \rrbracket)) \models \psi$, saying that an arbitrary relevant antecedent-world is a consequent-world. Following the above analysis of such epsilon-sentences, how likely we should take an arbitrary relevant antecedent-world to be a consequence-world is coordinated by an estimate of the fraction of consequent-worlds among the relevant antecedent-worlds. The more relevant antecedent-worlds we take to be consequent-worlds, the more likely should we take an arbitrary antecedent-world to be a consequent-world.

To see the advocated explanation at work, let us quickly go back to some of our original examples:

(1) If a given coin had been tossed, it would have landed heads,

(2) If I had bought a lottery ticket, I would have lost,

(3) If McCain had won, Obama would have lost.

We found that we do not rule out the first counterfactual outright. Since some of the relevant toss-worlds are heads-worlds, we should not rule out that an arbitrary toss-world is a heads-world. That we take the second counterfactual to be probable is because we should have a high credence that an arbitrarily selected antecedent-world is a consequent-world given that most antecedent-worlds are consequent-worlds. Finally, the last counterfactual is such that all relevant antecedent-worlds are consequent-worlds. In this case, we can be certain that an arbitrary antecedent-world is a consequent-world.

It seems, then, that the semantics in terms of the epsilon-operator contains the right ingredients to account for the subjective probabilities of counterfactuals. It may not have gone unnoticed, though, that the explanation of our credences in counterfactuals that I have given presupposes that counterfactual chances are uniformly distributed over the relevant antecedent-worlds. What happens when this assumption is relaxed? A second, perhaps less immediate question concerns the formal details of the interaction between probability and counterfactuals. If we add a subjective probability function to the picture, can we show that it accords with the evaluation constraint? The next two subsections focus on these questions.

6.2.1 Uniformity Reconsidered

On the present analysis, the subjective probabilities of counterfactuals estimate whether an antecedent-world which is chosen according to a uniform probability distribution is a consequent-world. This is because epsilon-terms have been described as selecting their referents in a completely unbiased fashion, not favouring any candidate world over another one. Thus, every relevant antecedent-world is given the same weight. This way of evaluating counterfactuals is in accordance with the evaluation constraint if and only if the counterfactual chances are taken to be uniform, that is if the uniformity constraint is adopted for counterfactual chances. To recall, the evaluation constraint requires that if the counterfactual chances are taken to be so-and-so, our credence in the corresponding counterfactual should be so-and-so. If and only if counterfactual chances are uniform, the counterfactual chances of the consequent under the antecedent equal the chances that an antecedent-world which is selected according to a uniform probability distribution is a consequent-world. For this reason, we will need to focus on the uniformity assumption in some more detail.

The uniformity assumption appears attractive partly because of its simplicity and partly because it provides a direct answer to the question of what the structure of counterfactual chances is. However, there are certain examples which challenge the uniformity assumption. Think, for instance, of a trick coin whose weight distribution is unbalanced so that it comes up heads with a probability of 2/3. Now, with respect to such a coin, consider the following counterfactual:

(4) If the coin had been tossed, it would have landed heads.

Assuming that the coin was not tossed and everything else is normal, it seems that our credence in this counterfactual should be about 2/3. It is not clear whether this credence can be explained if the uniformity assumption is adopted, for it seems that the distribution of heads-worlds among the relevant antecedent-worlds is only about 50%. But on this assumption, we would then have to assign a credence of 0.5 to the counterfactual which does not seem right.

Why would it be plausible to assume that the distribution of heads-worlds among the relevant antecedent-worlds is still about 50% despite the fact that the coin is not fair? The thought would be that the unbalanced weight distribution of the coin does not introduce three different possibilities two of which are such that the coin lands heads. This would result in a probability of 2/3 based on the uniformity assumption. Rather, it seems that there remain just two possibilities, namely one in which the coin lands heads and one in which it lands tails. Thus, the unusual weight distribution would not make a difference to the number of possibilities but only to the likelihood of these possibilities. One could, of course, try to resist this line of thought, perhaps by holding that worlds can still be different even if the same things happen in them. If so, then the account can stay as it is. But it would be nice not to be committed to this assumption, which is why we are going to seek for a way of doing without it.

The epsilon-operator has been construed as an operator which arbitrarily selects an object from the domain of the predicate it is attached to. A selection which is completely arbitrary seems to be a selection according to a uniform probability distribution.[20] So, we may see the epsilon-operator as implicitly indexed with a uniform probability function P. When the epsilon-operator is attached to a predicate, it represents an object which is selected from the extension E of the predicate according to the probability distribution $P(\bullet|E)$, that is with the probability P restricted to the extension of the predicate. So, if the total domain consists of 8 objects and there are four objects in the domain of the predicate, then attaching the epsilon-operator to the predicate denotes an object which has been selected with probability 1/4 (and not 1/8).

[20] As indicated earlier, there are certain domains over which there is no uniform distribution. In such cases, it is not clear what to say about the structure of the epistemic profile induced by the epsilon-operator, for a non-uniform distribution does not seem right either. I have to leave this problem open.

Once we describe the epsilon-operator this way, it is only a small step to recognize it as a special case of a more general conception. There is no need to start with a uniform distribution P. Instead, we could allow for probability distributions which assign different weights to the objects in the total domain. Then applying the epsilon-operator to a predicate would represent an object in the domain of the predicate which has been selected by a weighted choice process. The epsilon-operator as construed so far would be a special case when P is a uniform distribution. An epsilon-term such as $\ulcorner \epsilon_P x F(x) \urcorner$ would represent an object in the extension E of F which has been selected from E according to the probability distribution $P(\bullet|E)$. It does not seem to be much of an additional cost to accept the use of epsilon-terms which represent weighted choices.

Let us now apply the present idea to the case which is our main concern. The relevant domain for the epsilon-operator is the underlying set of possible worlds. The evaluation constraint makes use of a probability distribution Ch over this set of possible worlds. The counterfactual chance of B under A is then the probability $\mathrm{Ch}^A_w(B)$, that is the chance of B conditionalized on the set of relevant A-worlds. This chance distribution comes from Ch by conditionalizing on the set of relevant antecedent worlds, i.e. it is defined as $\mathrm{Ch}(B|f_w(A))$. We may now consider epsilon-operators indexed with such a chance distribution. They will represent antecedent-worlds selected according to the true counterfactual chance distribution of the world w. For instance, an epsilon-term like $\ulcorner \epsilon_{\mathrm{Ch}}\, w'(w' \in f_w(A))) \urcorner$ will represent a world selected from the set $f_w(A)$ with probability $\mathrm{Ch}(\bullet|f_w(A))$. Based on this, we can extend the semantics for counterfactuals in the following way:

Arbitrary Truth Conditions Extended. Let w be any world and $\ulcorner \phi > \psi \urcorner$ a counterfactual. Then $w \models \ulcorner \phi > \psi \urcorner$ iff

$$(\exists w' \in f_w([\![\phi]\!])) \supset (\epsilon_{\mathrm{Ch}}\, w'(w' \in f_w([\![\phi]\!])) \models \psi).$$

Thus, in the most interesting cases, a counterfactual is true at a world of evaluation w if the consequent is true at a relevant antecedent-world selected according to the counterfactual chance distribution of w. It is clear that this resolves the present issue. On the assumption that the counterfactual chances of B under A is x, we should think it x-likely that the selected antecedent-world is a consequent-world, for the antecedent-world is represented as being selected according to the counterfactual chances under A. Thus, by building the counterfactual chance distribution into the epsilon-operator, the present semantics can now explain the evaluation constraint even in cases in which the counterfactual chance distribution is not uniform.

6.2.2 Modelling the Epistemic Space

Although our semantics is formal, the explanation of how subjective probabilities attach to counterfactuals has been somewhat informal. We have derived the epistemic profile of counterfactuals from the epistemic profile of epsilon-sentences. Epsilon-terms present their referents under a veil of ignorance: we can know only with certain probabilities to which object they refer. This feature is key for using them to explain our uncertainty about counterfactuals. Can it be formally modelled? Can we give, for instance, a toy model in which the probabilities of counterfactuals can be seen as the probabilities of certain sets of worlds at which the epsilon-sentences are true? This task proves to be much more difficult than it may initially seem. In this section, I explain what difficulties one is confronted with and briefly indicate how one may try to find one's way around them.

To keep things simple, let us focus on a particular case. Suppose our space of metaphysical possibilities consists of a hundred worlds, $w_1, \ldots, w_{100}$. Assume further that we have already ruled out all worlds but the first nine, $w_1, \ldots, w_9$. These comprise the domain of our epistemic possibilities over which our credences are defined. For simplicity, we may assume that our credence function Cr is uniform, giving each world the same weight. Further, let us choose a coin-toss counterfactual as our main example. We may assume that there is no uncertainty over the relevant toss-worlds. So, to all of the nine worlds w_i the selection function f assigns the same set of relevant antecedent-worlds. Assume that these are the ten worlds $w_{20}, \ldots, w_{29}$, of which the first five are heads-worlds and the last five are tails-worlds. Then our evaluation constraint would predict a credence of 1/2 in the counterfactual 'If the coin had been tossed, it would have landed heads' ($\ulcorner A > B \urcorner$). If we now apply our semantics and look at the worlds at which the counterfactual comes out true, i.e. at the set $\{w : w \models \ulcorner A > B \urcorner\}$, can we show that it is assigned credence 1/2 by the function Cr? The answer is straightforwardly negative. There are two ways of seeing this. Let me explain them in turn.

Given that we have assumed that there is no uncertainty about which worlds are relevant—for all worlds $w_1, \ldots, w_9$ it is the set

$M := \{w_{20}, \ldots, w_{29}\}$—the epsilon-operator will select the same arbitrary world out of M relative to all our epistemic possibilities. Assume it selects w_{23}. This is a heads world and so the counterfactual would be true. But it would then be true at all our epistemic possibilities, for there is no uncertainty concerning the set of relevant worlds, and hence our credence would turn out to be 1. Similarly, if a tails-worlds had been selected, say w_{27}, the counterfactual would have been false, but it would have been false at all epistemic possibilities, and so a credence of 0 would have been predicted. No matter which world the epsilon-operator selects, the resulting proposition is one we are not uncertain about.

There is a more general way of seeing the problem. Simply note that no proposition is assigned credence $1/2$ by our function *Cr*. As the probability is uniformly distributed over nine worlds, any proposition is assigned a credence of $n/9$, with n ranging from 0 to 9. But because 9 is not divisible by 2, for no such n we have $n/9 = 1/2$. So, no matter which proposition is expressed by the counterfactual, it is not going to receive credence 1/2.[21]

What has gone wrong? I will suggest that the problem comes from equating the modal profile of a sentence with its epistemic profile. It is clear that the two can sometimes come apart. Take the sentence 'Hesperus is Phosphorus'. This sentence is metaphysically necessary and so its modal profile can be adequately represented by the set of all possible worlds. However, the sentence is not epistemically necessary—it is not irrational to assign to it a credence less than 1 under certain circumstances. Hence, its epistemic profile should not be represented by the set of all possible worlds. Conversely, consider a sentence like 'There is a thinker'. It may be epistemically necessary but it is clearly not metaphysically necessary. Moreover, the guise under which a modal profile is presented can make for epistemic differences. The sentence 'Hesperus is Hesperus' has a different epistemic profile than 'Hesperus is Phosphorus', although the two sentences have the same modal profile (they are both necessary). With this in mind, it is an immediate concern that something similar might happen when epsilon-terms are involved. After all, they have quite a special sense: they present their referents by only partially describing them and leaving what is left open to chance. We may therefore suspect that

[21] Effectively, we have illustrated that a triviality result for indicative conditionals based on finite domains and proven in Hájek and Hall (1994: 91) extends to counterfactuals on the present evaluation constraint.

our model above was too coarse-grained to capture the epistemic profile of epsilon-sentences.

What could a better model look like? Take world w_1, which is an epistemic possibility, and assume that as a matter of fact, the counterfactual is true at w_1, i.e. a heads-world has been selected relative to w_1. Can we therefore, from the perspective of w_1, be certain that the counterfactual is true? Not really, it seems, for there is no way of knowing this. From the perspective of w_1, the counterfactual might be true but it might equally well be false. The reason is that the epsilon-term does not reveal whether a heads-world or a tails-world has been selected. So, from the perspective of w_1, there are really two epistemic possibilities: one in which a heads-world gets selected and one in which a tails-world gets selected. If we want to model the epistemic space accordingly, we should pair the metaphysical possibility w_1 with two epistemically possible worlds w_1^1 and w_1^2. The same should be done to our credence. If w_1 deserves a credence of 1/9, then this bit of probability should be redistributed over w_1^1 and w_1^2 according to how likely they are from the perspective of w_1. In our sample case, they are both equally likely because the counterfactual chance of a heads-world and a tails-world are each 1/2. Consequently, the two epistemic worlds should each be assigned probability $1/9 \cdot 1/2 = 1/18$. If we do this for all our original nine metaphysical worlds, we arrive at an epistemic space consisting of the 18 worlds $w_1^1, w_1^2, \ldots, w_9^1, w_9^2$. In exactly half of them is the counterfactual true and given that the probability distribution remains uniform, our counterfactual will receive a credence of 1/2 as desired.

We could go on and push the construction further. The first thing to note is perhaps that in order to account for all possible consequents a counterfactual with the antecedent 'if the coin had been tossed' might have, there are more than two epistemic possibilities emerging from the perspective of w_1. To any world in M—the set of relevant antecedent-worlds relative to w_1—there is the possibility that this world gets selected. So, more realistically, we should represent the epistemic possibilities relative to w_1 by ten worlds $w_1^{20}, \ldots, w_1^{29}$ corresponding to the ten worlds in M. This will not alter the probability of the coin-toss counterfactual, for in half of them the counterfactual will be true, but it will now also allow us to calculate the subjective probability of other counterfactuals with the same antecedent. The next step would be to incorporate counterfactuals with a different antecedent. Here we could simply repeat the procedure on the already constructed space. So, step by step, we would split our starting worlds up into more and more epistemic possibilities corresponding to the possible selections of the epsilon-operator. Each time a world gets split up, its probability is redistributed on the newly emerging worlds with the weights specified by the counterfactual chances. If the counterfactual chances are uniform as in our example, each new world gets the same share, but if necessary, we can redistribute the probability with the weights assigned by the counterfactual chances. Ultimately, then, we arrive at an epistemic space which represents all the epistemic possibilities introduced by (non-nested) counterfactuals. (In much more generality, constructions of this kind have been studied by van Fraassen (1976) and more recently by Bacon (2015) in the context of satisfying a version of Adams's thesis for indicative conditionals. I have to leave it as an open question how the techniques they develop can be adapted to the present case.[22])

In conclusion, an expansion of the domain of possibilities seems to be what is required for constructing an adequate epistemic space in which to represent the subjective probabilities of counterfactuals. In a way, this reflects the idea that counterfactuals, and conditionals more generally, allow us to draw new distinctions we could otherwise not make (cf. Stalnaker 1984). We will come back to this issue in chapter 8, for it is closely related to possible triviality results for counterfactuals.

6.3 Knowability and Assertability

The present semantics for counterfactuals predicts a certain limitation of our epistemic access to counterfactuals. As we have seen, we can only know an arbitrary *F* to be *G* if all *F*s are *G*s. If only some *F*s are *G*, we cannot know whether an arbitrarily selected *F* is *G*, for we cannot know which object has been selected. By the specification of truth conditions in terms of the epsilon-operator, it follows then that we can—special cases aside—only know that a counterfactual is true if all relevant

[22] I should also admit that I have described the construction without any possible coordination constraints in mind. It should be noted that neither van Fraassen nor Bacon attempt to satisfy the substitution principle (Bacon argues explicitly against it), so it is not unlikely that if coordination constraints are imposed, further complexities will arise.

antecedent-worlds are consequent-worlds. We may therefore conjecture the following constraint to hold:

Unknown Counterfactuals. If not for all w' in $f_w(\llbracket\phi\rrbracket) : w' \models \psi$, then the counterfactual $\phi > \psi$ is in w not known to be true.

The phrase 'is not known' is to be understood as 'is not known by anyone at any time'. Note that this does not generate a modal claim: the constraint merely states that certain counterfactuals are not known in certain worlds. So, these counterfactuals need not be unknowable in the strong sense of it being metaphysically impossible to know them. We will return to this issue shortly.

An important point is not captured by the present constraint. To see this, note that counterfactuals would satisfy the constraint even if they had the standard truth conditions. However, this would be so because the counterfactual would be false under the specified conditions and for this reason unknown. The situation is interestingly different with respect to the truth conditions in terms of the epsilon-operator. A counterfactual may be true at a world w because the arbitrary antecedent-world is a consequent-world despite the fact that not all relevant antecedent-worlds are consequent-worlds. In such a case, the counterfactual will be true but unknown.

Structurally, we can identify this epistemic feature by noticing that under certain circumstances neither the counterfactual nor its negation can be known. Since either the counterfactual or its negation is true, this gives us cases in which true counterfactuals are not known. Let us call a sentence *undecided* iff neither it nor its negation is known to be true. Then we would have the following principle:

Undecided Counterfactuals. If neither for all w' in $f_w(\llbracket\phi\rrbracket) : w' \models \psi$ nor for all w' in $f_w(\llbracket\phi\rrbracket) : w' \models \neg\psi$, then the counterfactual $\ulcorner\phi > \psi\urcorner$ is undecided in w.

In words: if neither all antecedent-worlds are consequent-worlds nor all antecedent-worlds are non-consequent-worlds, then a counterfactual is undecided. This is a direct consequence of our semantics and the constraint on counterfactual knowledge. If the first condition holds, then the counterfactual $\ulcorner\phi > \psi\urcorner$ is not known and if the second condition holds, then the counterfactual $\ulcorner\phi > \neg\psi\urcorner$ is not known. But on the present semantics, $\ulcorner\neg(\phi > \psi)\urcorner$ logically implies $\ulcorner\phi > \neg\psi\urcorner$: a negation of a counterfactual can only be true if the antecedent is not impossible and the arbitrary antecedent-world is not a consequent-world which makes it a non-consequent-world (this is related to the fact that conditional excluded middle holds on the present semantics; see section 7.3 for discussion). Given this, it is clear that we cannot know in the specified situation that $\ulcorner\neg(\phi > \psi)\urcorner$ is true, for otherwise we could infer $\ulcorner\phi > \neg\psi\urcorner$ which we cannot know. So, neither $\ulcorner\phi > \psi\urcorner$ nor $\ulcorner\neg(\phi > \psi)\urcorner$ will be known. But, of course, one of the two sentences will be true, although we cannot know which. This, then, captures the main epistemic consequence of the present semantics: there are true counterfactuals which are not known for semantic reasons.[23]

Such a limitation of our epistemic access to counterfactuals squares well with the amount of counterfactual knowledge we are willing to ascribe to ourselves. A paradigm example of a counterfactual which, in typical circumstances, turns out to be undecided is a counterfactual like 'If the coin had been tossed, it would have landed heads'. We do not take ourselves to know that this counterfactual is true. More importantly, we do not take ourselves to know that this counterfactual is false either as one might expect on the standard account. Rather, we are happy to say that we do not know *whether* the coin would have landed heads if it had been tossed.[24] This intuition can also be expressed by saying that there is no way of telling how the coin would have landed. Similarly in the other cases we have discussed.

[23] The present principles are not without exceptions if they are taken to apply to counterfactuals which contain the epsilon-operator. A counterfactual of the form 'If ϕ had been true, then the world would have been exactly like the arbitrarily selected relevant ϕ-world' can be known without all relevant antecedent-worlds verifying the consequent. Such counterexamples come about because the consequent of the counterfactual states something which can be known a priori to be satisfied by the arbitrary antecedent-world. Thanks to Timothy Williamson for pointing this out to me.

[24] I take this to be an intuitive datum of moderate strength. It is clear that proponents of the standard account or error theorists like Hájek (2007) will disagree, claiming that we *can* know that such counterfactuals are false. To defend their position, they will need an explanation of why we are pre-theoretically more inclined to take an agnostic stance. Others may explain our tendency to take an agnostic stance not by positing limited epistemic access, but by holding that there is no matter of fact which determines the counterfactuals' truth-value (they would be metaphysically indeterminate). My claim here is only that general lack of knowledge in such cases provides a good explanation, not that it is the only possible explanation.

Although the present account implies a certain kind of scepticism concerning certain counterfactuals, it should be noted that there is no obstacle from the present point of view that many counterfactuals can be known and are known to be true. In whichever situation the standard semantics predicts that a counterfactual is true because all relevant antecedent-worlds are consequent-worlds, the present account is compatible with the counterfactual being known. Since there are still many cases in which the standard truth conditions are satisfied, there are still many cases in which a counterfactual may be known.[25] To some extent, the standard truth conditions turn out to approximate the knowability conditions of counterfactuals.

Finally, let us briefly return to an issue mentioned earlier. As they stand, the principles concerning the epistemic access to counterfactual facts do not have any modal force. They only state that certain counterfactuals are not known under certain circumstances. Does the reasoning behind these principles suggest anything stronger? A stronger claim would be that the counterfactuals under consideration cannot be known in the strong sense of it being metaphysically impossible that they are known. But such a stronger claim is not warranted. The present semantics does not imply that there are true counterfactuals which are strictly unknowable. The reason why this is so is simple: nothing in the present semantics suggests that if for a world w neither all worlds in $f_w(A)$ are ψ-worlds nor all worlds in $f_w(A)$ are $\neg\psi$-worlds, then there is no world w' such that either all worlds in $f_{w'}(A)$ are ψ-worlds or all worlds in $f_{w'}(A)$ are $\neg\psi$-worlds. But if there is such a world w', then it is compatible with the present semantics that the corresponding counterfactual is known in w' despite the fact that it is not known in w. An example might help to illustrate this. Suppose in the actual world a given lottery is fair. Then the semantics suggests that the counterfactual 'If I had bought a lottery ticket, I would have lost' is unknown. This is, of course, compatible with there being a possible world in which the lottery is rigged such that all tickets are determined to lose. In such a world, I may know the counterfactual to be true by knowing of the manipulation. So, the present account does not predict that the counterfactuals which are not known for semantic reasons are also strictly unknowable.

[25] Exactly how many counterfactuals are knowable will depend on how broad or narrow one's concept of relevance is. If typically a larger class of antecedent-words is relevant, fewer counterfactuals will be knowable, and *vice versa*. Compare the discussion in section 3.9.

Let me make a few remarks on assertions of counterfactuals. Of course, what to say will to a large extent depend on a general outlook on assertion. To illustrate the situation with a particular example, I will here follow Williamson (1996, 2000: ch. 11) in assuming that the norm of assertion is knowledge: one should assert something only if one knows it. Given the constraint on counterfactual knowledge, we can derive a constraint on the assertability conditions of counterfactuals:

Assertability of Counterfactuals. If not for all w' in $f_w([\![\phi]\!]) : w' \models \psi$, then the counterfactual $\ulcorner \phi > \psi \urcorner$ is not assertable in w.

A counterfactual can only be known if all relevant antecedent-worlds are consequent-worlds, and a counterfactual should only be asserted if it is known to be true. Thus, the constraint follows from the knowability conditions of counterfactuals together with the general assumption that only what is known should be asserted.

Now, it seems to be a datum that in typical situations one should neither assert outright a counterfactual like 'If the coin had been tossed, it would have landed heads' nor 'If the coin had been tossed, it would have landed tails'. But on the present account, one of the two counterfactuals will be true. Initially, the fact that neither of the two counterfactuals is assertable may be levelled as an objection against the present theory, for it may be taken to suggest that neither of the two counterfactuals is true. By the principle above, however, the present datum can be explained in a straightforward way: the two counterfactuals should not be asserted because they are not known to be true.[26] Besides, it can also be explained why there is the feeling that the non-assertability of these counterfactuals is somewhat more robust than in other cases in which a sentence happens to be non-assertable in a given situation. This would be due to the fact that in typical situations our epistemic access to such counterfactuals is limited for principled reasons: if neither all antecedent-worlds are consequent-worlds nor all antecedent-worlds are non-consequent-worlds, there is nothing we can do to find out whether such a counterfactual is true.

[26] A further question concerns what one should say about the assertability of counterfactuals which have a 'high chance' consequent instead of one with middling chances. If one takes the former to be assertable, one will probably take issue with the knowledge rule for assertion, at least this is what one would have to do on the present semantics for counterfactuals.

6.4 Revisiting the Puzzle

The basic argument against truth conditions starts out with a puzzle concerning the subjective probabilities of counterfactuals. High subjective probability of a counterfactual seems to be compatible with low subjective probability that the standard truth conditions obtain, while knowledge of a counterfactual seems to put us in a position to know that the standard truth conditions obtain. We have already seen that the present semantics underwrites these two features concerning the epistemic profile of counterfactuals by showing that it gives rise to the evaluation constraint.

As already pointed out, the truth conditions in terms of the epsilon-operator are weaker than the standard truth conditions: if the standard truth conditions obtain, the present truth conditions will obtain as well but not always *vice versa*. So, from the perspective of the present semantics, there is no need to question the part of the puzzle which suggests that if counterfactuals have truth conditions, they must be weaker than the standard truth conditions. Instead, what one will need to cast doubt on is the claim that if counterfactuals have truth conditions, they must be at least as strong as the standard truth conditions.

As we have seen, this claim is based on the principle that if a proposition *p* does not imply a proposition *q*, then it is possible to know that *p* without being in a position to know that *q*. We also saw that this principle cannot hold in full generality. Not only is it falsified by unknowable truths, it is also called into question by propositions which can only be known to be true by knowing a slightly stronger proposition.

Now, on the present account, the proposition expressed by a counterfactual relates to the proposition it is taken to express according to the standard semantics in a particularly interesting way. Counterfactuals are construed as weaker than the standard truth conditions. But in order to know that a counterfactual is true, one needs to be in a position to know the slightly stronger proposition that the standard truth conditions obtain. For we can only know that an arbitrarily selected antecedent-world is a consequent-world if we are in a position to know that all antecedent-worlds are consequent-worlds. Thus, on the present account, the proposition expressed by a counterfactual and the proposition it is assumed to express on the standard semantics constitute a counterexample to the implication principle. Hence, if the puzzle is turned into an argument against truth conditions for counterfactuals, the step in which this implication principle is applied can be resisted.

Thus, we can win back truth conditions for counterfactuals by trading in a certain kind of epistemicism for counterfactuals: there are principled limits on our epistemic access to counterfactual facts. The puzzling credences in counterfactuals typically reflect this kind of limitation. Given that we have identified only the implication principle as a possible loophole in the argument against truth conditions, epistemicism seems to be without alternatives if the evaluation of counterfactuals is to be explained in terms of truth conditions.

6.5 Conclusion

On the present theory, counterfactuals involve an element of arbitrariness. They are true iff an arbitrarily selected relevant antecedent-world is a consequent-world. These truth conditions are weaker than the standard truth conditions, for they do not imply that all relevant antecedent-worlds are consequent-worlds. Nevertheless, given the element of arbitrariness, counterfactuals have a more demanding epistemic profile than their truth conditions might suggest. In order to know that a counterfactual is true, the standard truth conditions have to be satisfied. But in situations of uncertainty, less is required: that the truth conditions of counterfactuals are satisfied is still supported to a high degree if most but not all relevant antecedent-worlds are known to be consequent-worlds. This helps solve the puzzling evaluation of counterfactuals, which suggests that our credences go with an estimate of a possibly weighted fraction of the consequent-worlds among the relevant antecedent-worlds.

What we have not yet seen is how this theory fares with respect to other problems in the theory of counterfactuals which have little to do with probability. Problems of this kind will be the topic of the next chapter. I hope this will further sharpen the picture of counterfactuals suggested by the present theory.

7

Applications

The present semantics has various features which bear on longstanding issues in the debate about counterfactuals, like the duality thesis or conditional excluded middle. It is worth seeing how it resolves these issues. In some, perhaps even in most cases, the present semantics makes the somewhat more controversial prediction regarding these issues. For this reason, the following discussion is also intended as a partial defence of the epsilon-based semantics.

There are some questions which are missing from our discussion: the presuppositions of counterfactuals, Sobel sequences and disjunctive antecedents, to name just a few. This is not because I think that these issues are not important, it is rather because the account developed so far does not bear in any direct way on them and can be combined with various, but perhaps not all approaches to solve these problems.

7.1 Counterfactuals with a True Antecedent

On a semantics like the present one which binds the truth-value of a counterfactual to what is true at a single world, the validity of *modus ponens* has the consequence that the actual truth of antecedent and consequent is sufficient for the truth of the corresponding counterfactual. That is, the following principle ('CI' for *conjunction implication*) holds:

(CI) $((A \land B) \supset (A > B))$.

The reason is the following. Suppose A and $\ulcorner A > B \urcorner$ are true at the actual world, w. If modus ponens is valid, then B must be true at w, too. But this is only ensured if the relevant A-world on which the truth of $\ulcorner A > B \urcorner$ depends is w. If it were some other world w', B would be true at w' but could be false at w. In general, then, the truth of A at w will make w the relevant antecedent-world. Yet this implies that if A and B are true at any world w, the counterfactual $\ulcorner A > B \urcorner$ will be true at w, for its truth will depend only on whether B is true at w. So, on any kind of Stalnaker semantics, one cannot have modus ponens without having the conjunction implication principle. This is different in standard semantics: there one can have modus ponens without the conjunction principle by imposing weak centering while rejecting strong centering (cf. Lewis 1973: 29–31).

As we have seen, the easiest way of validating modus ponens on the present semantics is to adopt strong centering: if A is true at w, then $f_w(A) = \{w\}$—if the antecedent is true at the world of evaluation w, the set of relevant antecedent-worlds contains only w. In this case, the epsilon-operator is forced to select the world of evaluation for true antecedents. If antecedent and consequent are true at the world of evaluation, the consequent will thus be true at the selected world and so the counterfactual will be true, too. As was pointed out in section 6.1.3, an alternative way of validating modus ponens would be to only adopt weak centering for the selection function f, but to ensure via a coordination constraint on the selection of antecedent-worlds that the epsilon-operator has to select the world of evaluation if the antecedent is true at it.

To simplify the discussion, I will presuppose that modus ponens is valid for counterfactuals. This assumption is not trivial as one would like to check whether the potential counterexamples to modus ponens for indicative conditionals have interesting analogues in the counterfactual case.[1] In any event, it should be clear that modus ponens will hold in a wide range of cases. So we might as well, for the purposes of our discussion, pretend that the present semantics is committed to the universal validity of the conjunction principle.

Potential counterexamples to the conjunction principle typically consist of an antecedent A and a consequent B where there is (i) no connection between A and B or (ii) not the right kind of connection. An instance of (i) is proposed by Fine (1975: 453):

(1) If I were to raise my little finger, there would be rainfall this winter,

where we should suppose that the finger will indeed be raised and that there will be rainfall. Instances of (ii) concern cases where the antecedent

[1] Counterexamples to modus ponens are proposed in MacFarlane and Kolodny (2010) and McGee (1985).

made the consequent unlikely, did not make it likely enough or led to it only through an indeterministic process (the first example is adopted, with a twist, by Bennett 1974: 387f. in response to Lewis's 1973: 27 discussion of the same counterfactual; the second one is due to Fine 1975: 453, and the third one to Bennett 2003: 240; for further discussion, see Walters 2009):

(2) If Caspar had come, the party would have been a good one (the party was good, but Caspar's presence almost ruined it).

(3) If he had worked hard, he would have passed (he worked hard, but he passed only because he cheated, too).

(4) If the coin were tossed, it would fall heads.

In response to these kinds of cases, and in particular to the discussion in Bennett (2003: §92), Walters (2009) has given the conjunction principle an extensive defence (see also the reply by Ahmed 2011 and the response by Walters 2011). Walters observes that the truth of Morgenbesser counterfactuals would be hard to reconcile with failures of the conjunction principle. To recall, suppose, for instance, that a coin was tossed and came up heads. We might then say

(5) If I had bet on heads, I would have won.

But given that winning after betting on heads implies that the coin came up heads and that it must have been tossed before that, it seems incoherent to accept this counterfactual while denying that the coin would have come up heads if it had been tossed. If you take it not to be the case that the coin would have come up heads if it had been tossed, why think that you would have won if you had bet on heads? Walters develops this observation into a precise argument for the conjunction principle. However, his argument relies on principles like the substitution principle (see section 6.1.3) which are not part of the base logic resulting from the present semantics. So, although I am very sympathetic to the overall spirit of Walters's argument, it is not clear whether it can be adopted from the perspective of the present semantics.

Instead of giving a positive argument for the conjunction principle, I will only sketch what I take to be a promising line of defence against the potential counterexamples. The strategy will be to follow Lewis (1973: 27f.) in finding a pragmatic explanation of why these counterfactuals strike us as odd. I think there is some initial justification for this approach. It does not seem that we have clear intuitions that the candidate counterfactuals are false. We would not, I take it, go ahead and assert their negation. Still, it seems fair to say that there is something wrong with them, but that need not necessarily be falsity.

A first thought in search of a pragmatic explanation might be that with the conjunction principle in place, conjunctions are at least as strong as the corresponding counterfactuals. Even more, they are strictly stronger, for counterfactuals do not imply the corresponding conjunction. In the light of this, asserting a counterfactual merely on the basis of the conjunction means to assert something weaker where one could have asserted something stronger (cf. Lewis 1973: 28). Gricean maxims suggest that this is inappropriate. Unfortunately, it is unclear whether this kind of explanation is on the right track.[2] In situations in which we assert something weaker than we could have, it is typically possible to regain appropriateness by adding something like '. . . indeed [the stronger assertion]'. For example, after saying 'Mary is part of the committee' while knowing that she is actually the chairwoman, I could continue '. . . indeed, she is the chair of the committee'. Now consider one of the problematic counterfactuals followed-up with the corresponding conjunction: 'If the coin were tossed, it would fall heads, indeed, it will be tossed and it will fall heads'. It is somewhat unclear to me what one should say about this sentence. It sounds a bit funny, but I don't hear it as clearly infelicitous either. Tentatively, I conclude that the logical relation between counterfactuals and the corresponding conjunction may well play a role in a pragmatic explanation, but does not on its own provide the material for a full story.

Trying to exploit the fact that a conjunction is logically stronger than the corresponding counterfactual in giving a pragmatic explanation may seem to run contrary to the first impression about the potential counterexamples. At first sight, it seems that the counterfactuals communicate something—a relevant connection between antecedent and consequent perhaps—which goes beyond the corresponding conjunction (rather than the other way around). Proponents of the counterexamples take this to suggest that counterfactuals are not implied by the corresponding conjunction. But one could explore the possibility that the added elements

[2] Thanks to Alan Hájek for helping me with this problem.

of information are only pragmatically conveyed rather than semantically expressed. The basic idea would be that as hearers, we make inferences about the grounds on which a counterfactual is asserted. If the context does not provide any specific clues, we may assume that the grounds are typical or normal. Normal grounds for asserting, for instance, that the party would have been a good one if Caspar had come or that he would have passed if he had worked hard are ones where Caspar's presence boosts the party or where his working hard was the only reason for his success. If the grounds for asserting a counterfactual are not normal as it is the case when they merely consist in holding the conjunction, then asserting the counterfactual may be misleading unless it is a salient feature of the context that the grounds are different from what they normally are.

A third, not necessarily independent aspect which is likely to bear on the present issue has to do with the presuppositions of counterfactuals. It is generally agreed that counterfactuals do not presuppose (nor semantically imply) the falsity of the antecedent (see the cases discussed in Anderson 1951 and Stalnaker 1975: 145f.). Nevertheless, asserting a counterfactual seems only appropriate, at least in typical situations, if it is an epistemic possibility in the context that the antecedent is false (cf. von Fintel 1999). If it is common ground between all parties of the conversation that the antecedent is true, then using a counterfactual seems inappropriate (to confirm, consider 'We all agree that Jacky was in the bank. # If she had been in the bank, . . .'). This hypothesis is compatible with the observation by Anderson and Stalnaker that a speaker may use a counterfactual in order to argue for the truth of the antecedent, for in these cases the addressees do not yet grant, or are not assumed to grant, the truth of the antecedent.

If these three factors all play a role, then contexts in which the problematic counterfactuals are fine have to meet at least three conditions: (i) the speaker should not know that the conjunction is true or else the context should not require her to be maximally informative, (ii) the context should make it clear that the counterfactual is asserted on non-normal grounds and (iii) the truth of the antecedent should not be common ground. It is not clear whether contexts of this kind can always be found. But here is a way of doing this for (1), suggested to me by Alan Hájek: 'There will be rainfall this winter whatever I do. If I were to pray for a drought, there would be rain. If I were to gather all the world's leading scientists to try to prevent rain, there would be rain. If I were to raise my little finger, there would be rain . . .'. In this context, the problematic counterfactual sounds absolutely fine. Part of the reason is that the context makes it clear that the counterfactual is not asserted due to a relevant connection between antecedent and consequent, but rather because of a lack thereof. It is harder to find a similar context for a coin toss case because we typically do not have knowledge about future coin tosses as long as the underlying process is chancy. But suppose I say: 'I just talked to a time traveller, though I am going to tell you only this: If the coin were tossed, it would fall heads'. Again, the counterfactual seems fine.

As far as I can see, one has to be even more creative regarding the remaining two problem cases. Let us try to do it for the third case only. To recall, a candidate worked hard and passed, but he passed only because he cheated, too. Can we embed this scenario into a context to make 'If he had worked hard, he would have passed' appropriate? Suppose I am the secretary at a philosophy department. I know the results of the exam, but I take my obligation not to tell anyone very seriously, though I do not mind giving hints. A friend of the candidate is worried that he may not have passed because one may have found out about his cheating or perhaps cheating was not even enough and he should have worked harder, for he is always so lazy. Now I say: 'I cannot tell you whether he passed. But I can tell you this: If he had worked hard, he would have passed'. Admittedly, it is a somewhat curious thing to say, but as far as pragmatics are concerned, it strikes me as felicitous. Given the severe constraints on cases for which we would expect such counterfactuals to be fine, it should be no wonder that they are rare and require a fair amount of creativity to be set up.

7.2 Duality

A prominent hypothesis, put forward by Lewis (1973: ch. 1.5), is that 'would'-counterfactuals and 'might'-counterfactuals are duals of each other. The duality thesis can be given by the following schema (the arrow '$\diamondsuit\!\!\rightarrow$' represents 'might'-counterfactuals):

Duality. $(\phi > \psi) \equiv \neg(\phi \diamondsuit\!\!\rightarrow \neg\psi)$.

For instance, the following two counterfactuals would be equivalent:

(6) If Terry had scored, Chelsea would have won.

(7) It is not the case that if Terry had scored, Chelsea might not have won.

By exchanging the negation sign, the truth of a 'might'-counterfactual is supposed to imply the falsity of the corresponding 'would'-counterfactual:

(8) If the coin had been tossed, it might have landed tails.

implies, by duality, the falsity of the counterfactual

(9) If the coin had been tossed, it would have landed heads.

This example also shows that the duality thesis fails on the present account. It is clear that (a) if the coin had been tossed, it might have landed tails and (b) if the coin had been tossed, it might have landed heads. But our semantics has it that either (a) if the coin had been tossed, it would have landed heads or (b) if the coin had been tossed, it would have landed tails, for the arbitrarily selected antecedent-world is either a heads-world or a tails-world. No matter which world gets selected, either the (a)-pair of a 'might'-counterfactual and a 'would'-counterfactual or the corresponding (b)-pair constitutes a counterexample to the duality thesis.

Yet the duality thesis is not unproblematic. Consider the following sentences, which draw on Stalnaker (1978: 100f.):

(10) If I had bought a lottery ticket, I might have won, but I am fairly sure that if I had bought a lottery ticket, I would not have won.

(11) If I had bought a lottery ticket, I might have won, but it is likely that if I had bought a lottery ticket, I would not have won.

These sentences are clearly assertable in appropriate circumstances. But this should not be so if the duality thesis were true, for it construes the two embedded counterfactuals as logically equivalent to the negations of each other. But instances of a schema like 'p, but it is likely that $\neg p$' are, for reasons akin to Moore's paradox, not assertable. This is evidence against duality.[3]

[3] A possible comeback for a defender of the duality thesis would be to explain the plausibility of the two sample sentences by reconstructing the epistemic phrases as having narrow scope only. For problems with this strategy, see section 1.8.

Even if the duality thesis is assumed to fail, it would be nice to explain why the duality thesis appears attractive. For a start, it does not seem difficult to account for 'might'-counterfactuals in the present framework. Since the semantics is based on a function selecting the set of relevant antecedent-worlds, we can, for example, incorporate the standard approach to 'might'-counterfactuals: a 'might'-counterfactual of the form $\ulcorner\phi \diamondto \psi\urcorner$ is true iff ψ is true in at least one relevant ϕ-world.[4] More formally:

Counterfactual Possibility. $w \models (\phi \diamondto \psi)$ iff $\exists w' \in f_w(\llbracket\phi\rrbracket) : w' \models \psi$.[5]

Of course, the standard truth conditions are dual to the truth conditions for 'might'-counterfactuals so defined. Recall from section 4.1 that '$\boxright$' was simply defined to have the standard truth conditions without any claim that it corresponds to natural language counterfactuals. On the present picture, it is naturally seen as expressing counterfactual necessity:

Counterfactual Necessity. $w \models (\phi \boxright \psi)$ iff $\forall w' \in f_w(\llbracket\phi\rrbracket) : w' \models \psi$.

We may ask whether the box-arrow can be expressed in English. As far as I know, there are no 'must'-counterfactuals in English, where one would simply substitute a 'must' for the 'might' in a 'might'-counterfactual.[6] Perhaps, the following construction with 'have to' comes closest to the box-arrow in English:

[4] On the standard account, a 'might'-counterfactual is not treated as a 'would'-counterfactual the consequent of which is governed by 'might' with a higher scoping 'would' left implicit. See Swanson (2010: 537) for discussion of this latter possibility. Cf. also Geurts (2004) and Leslie (2009). It may be, however, that the latter analysis would result in the same truth conditions for simple 'might'-counterfactuals. Some constraints on which worlds are accessible for the modal operator would need to be imposed, for it is implausible to assume that a 'might'-counterfactual expresses that something would have been metaphysically possible if the antecedent had been true. It may then be that the set of relevant antecedent-worlds is the best candidate for providing the set of accessible worlds.

[5] On this account, a 'might'-counterfactual is false if its antecedent is impossible, but this effect could be removed by imposing an appropriate proviso. Another issue concerns the centering conditions: If strong centering is adopted for the selection function f in order to account for 'would'-counterfactuals, one could consider lifting this requirement in accounting for 'might'-counterfactuals.

[6] This may have a simple explanation: if the hypothesis mentioned in the introduction is correct that the governing modal in the consequent of a counterfactual always assumes a past tense form such as 'would' or 'might' (the past tense of 'will' and 'may'), then there are no 'must'-counterfactuals in English because 'must' does not have a past tense.

(12) If Terry had scored, Chelsea would have had to win.

Such counterfactuals seem to be judged false if at least one of the relevant antecedent-worlds is not a consequent-world. For instance, it is clear that the following counterfactual is false:

(13) If the coin had been tossed, it would have had to land heads.

Since the coin might have landed tails, it seems to follow that it would not have had to land heads. Conversely, if it would not have had to land heads, it might have landed tails. Whether counterfactual necessity can really be expressed by 'have to'-counterfactuals is something I would like to leave open.[7] But for the purposes of this section, I will tentatively assume that they can be.

As defined, all three kinds of counterfactual statements operate semantically on the same set of relevant worlds. One is accounted for by a universal quantifier, the other by the variable binding epsilon-operator, and the last one by an existential quantifier. This can account for logical relations between them. Setting the case of an impossible antecedent aside, the following three counterfactuals turn out to be of descending strength:

(14) If the coin had been tossed, it would have had to land heads.

(15) If the coin had been tossed, it would have landed heads.

(16) If the coin had been tossed, it might have landed heads.

Thus, 'would'-counterfactuals are construed as being of intermediate strength: they are sandwiched between counterfactual necessity and counterfactual possibility.

The descending strength of the three types of counterfactuals is modelled in close parallel to the variable strength of statements about the future. Compare:

(17) The coin has to land heads.

(18) The coin will land heads.

(19) The coin might land heads.

[7] Clearly, the English sentences are 'would'-counterfactuals whose consequent is modalized. So, the suggested semantics is not faithful to the most likely syntactic structure of these sentences. Nevertheless, the assigned truth conditions may still be correct (cf. footnote 4).

If we focus on an objective reading of 'might' and 'has to', the three sentences seem to be of descending strength in a way similar to the counterfactuals above. To a first approximation, the first sentence is true iff in all presently possible futures the coin lands heads, the second is true iff in the actual future the coin lands heads, and the last one is true iff in some presently possible future the coin lands heads. This parallels our semantics of the corresponding counterfactuals. Of course, the parallel can only go so far: there is a sense in which sentences about the future have a day of reckoning, while 'would'-sentences in the scope of a counterfactual antecedent do not. The parallel is restored by arbitrarily picking a counterfactual future.

In fact, this may be a fairly close parallel. As briefly mentioned in the introductory chapter, it is a common linguistic hypothesis that 'would' is a certain kind of past tense form of 'will'. Given this, one would expect that 'would' inherits the modal character of its origin 'will'. Since 'will'-sentences express propositions of intermediate strength about the future, it is much more natural to assume that 'would'-sentences express intermediate propositions as well rather than that they express a certain kind of counterfactual necessity as the standard account predicts. Alternatively, it may be possible to treat 'will' as a necessity operator operating on the set of all possible futures. It then would make sense to treat 'would' in a similar way. However, I take it to be the more plausible option that 'will'-sentences are not expressions of future necessity.[8]

With these considerations in mind, let us get back to the question of why the duality thesis appears attractive. Here we may note an interesting phenomenon. Although a 'might'-counterfactual and a corresponding 'would'-counterfactual with the negation signs filled in can both be true, the present account has it that they cannot both be known to be true. For, if the relevant 'might'-counterfactual is true, there will be among the relevant antecedent-worlds a world at which the consequent of the 'would'-counterfactual is false. By the knowability constraint

[8] Note that philosophers such as Thomason (1970) or MacFarlane (2003) who advocate a non-standard account of future contingents actually endorse a non-classical view. A sentence of the form ⌜Will ϕ⌝ is said to be true iff ϕ is true at all possible futures, false iff ϕ is true at no possible future and neither true nor false otherwise. The clause for truth may look as if they treat 'will' as a necessity operator, but the clause for falsity makes it evident that this is not the case, for a statement of necessity would be false if the relevant sentence is false at *some* possible future rather than being false at *all* possible futures.

(see section 6.3), the 'would'-counterfactual can therefore not be known. Hence, the conjunction with the 'might'-counterfactual cannot be known and would, by the knowledge account of assertion, not be assertable. In effect, this explanation deems sentences like

(20) If Terry had scored, Chelsea would have won, but Chelsea might (still) have lost (even) if Terry had scored.

to be Moore-paradoxical: they are consistent, but the second conjunct undermines the possibility of knowing the first one. Thus, although duality fails on the semantic level, it has a true pragmatic projection: the relevant pairs of 'might'-counterfactuals and 'would'-counterfactuals are not co-assertable.[9]

One application of the failure of duality concerns the recent debate about counterfactuals and objective chance.[10] On the assumption of non-trivial objective chances, many 'might'-counterfactuals seem true. By duality, many 'would'-counterfactuals would come out as false. For instance, there being a small objective chance that q conditional on p seems to support the claim that if it had been the case that p, it might have been the case that q. Given a slight objective chance that the plate flies off sideways, it seems true that

(21) If I had dropped the plate, it might not have been broken.

By duality, we would have to infer the falsity of

(22) If I had dropped the plate, it would have been broken.

In an indeterministic world, many cases have the underlying structure of this example, which would force many 'would'-counterfactuals to be

[9] DeRose (1999) and Stalnaker (1978) also give pragmatic explanations of the phenomenon (for critical discussion, see Hájek 2007). However, their explanations differ from the present one in that they take 'might'-counterfactuals to be epistemic or quasi-epistemic expressing something like the epistemic possibility of the corresponding 'would'-counterfactual. See also the discussion of *weak counterfactual scepticism* in Eagle (ms). In contrast, 'might'-counterfactuals can be construed as non-epistemic or ontic on the present semantics.

[10] See Hájek (2007), Hawthorne (2005), and Williams (2008). Many thanks to Tobias Rosefeldt for drawing my attention to this application.

false.[11] In this way, the duality thesis threatens many counterfactuals with falsity.[12] On the present account, there is no need to go to that extreme. Since the duality thesis fails, the truth of many 'might'-counterfactuals does not imply the falsity of many 'would'-counterfactuals. Although the plate might have flown off sideways, we can still be fairly confident that it would have been broken.

What we have seen is that even if many 'might'-counterfactuals are true, the corresponding 'would'-counterfactuals need not be false. In showing this, I went along with the assumption, endorsed in Hájek (2007) and considered favourably in Hawthorne (2005), that a positive conditional physical chance implies the corresponding 'might'-counterfactual. Let me just add that this is an assumption which is not implied by the present account, although it could be added to it. What is a consequence of the present account is that the truth of a 'might'-counterfactual is implied by there being a positive counterfactual chance of the consequent under the antecedent. But a positive physical chance need not be taken to always imply a positive counterfactual chance. Hence, it is compatible with the present account that a plate flying off sideways is just too remote a possibility for being included in the set of relevant antecedent-worlds (cf. Williams 2008 in response to Hawthorne 2005). On such a conception, we might even be able to know, and therefore be able to assert, that the plate would not have flown off sideways if it had been dropped (cf. also section 3.9).

7.3 Conditional Excluded Middle

On the present account, there is a certain kind of flexibility concerning the logic of counterfactuals. By adding or omitting constraints on the selection function f and possibly also through what I have called coordination constraints, the logic of counterfactuals can be influenced.

[11] It is not quite clear to what extent the assumption of indeterminism is essential to this argument. Even in a deterministic world, many (but perhaps less) 'might'-counterfactuals of the envisaged kind still seem to be true. Cf. Hájek (2007).

[12] Hájek (2007) embraces this argument. In the postscript to Lewis (1979), Lewis suggests resolving the conflict by positing an ambiguity in 'might'-counterfactuals. On the reading which is captured by the duality thesis, the obtaining of the relevant objective chances would not, according to Lewis, make the counterfactual true. On a second 'would-be-possible'-reading, the 'might'-counterfactual is taken to be true without obeying duality and therefore would not force a corresponding 'would'-counterfactual to be false.

However, there is also a limit to this flexibility: certain inferences will be valid even if no constraints are imposed. So, there is a certain *base logic* for counterfactuals which results already from the general shape of the present semantics. There is one important consequence of this base logic which is controversial in the debate about counterfactuals: the validity of *conditional excluded middle.*

Conditional Excluded Middle. $(\phi > \psi) \vee (\phi > \neg\psi)$.

In words: if *A* had been the case, *B* would have been the case, or, if *A* had been the case, *B* would not have been the case. This principle is valid on the present semantics: if the antecedent is impossible, then both counterfactuals are vacuously true and if the antecedent is not impossible, then the selected ϕ-world will either be a ψ-world or a $\neg\psi$-world, in which case either $\ulcorner(\phi > \psi)\urcorner$ or $\ulcorner(\phi > \neg\psi)\urcorner$ will be true. In contrast, conditional excluded middle is invalid according to standard semantics, for it may be that neither all relevant ϕ-worlds are ψ-worlds nor all ϕ-worlds are $\neg\psi$-worlds.

Lewis (1973: 79) calls this principle 'the principal virtue and the principal vice of Stalnaker's theory', but he admits that denying it sounds incoherent:

> Given conditional excluded middle, we cannot truly say such things as this:
>
> *It is not the case that if Bizet and Verdi were compatriots, Bizet would be Italian; and it is not the case that if Bizet and Verdi were compatriots, Bizet would not be Italian; nevertheless, if Bizet and Verdi were compatriots, Bizet either would or would not be Italian.*
>
> [...] I want to say this, and think it is probably true; [...] But off hand, I must admit, it does sound like a contradiction. (Lewis 1973: 80)

This indicates that conditional excluded middle enjoys a certain amount of intuitive plausibility.[13] But I do not want to play this card too strongly, for an opponent of conditional excluded middle might explain its plausibility as resulting from scope confusions as Williamson (1988) does: instances of conditional excluded middle might be confused with

[13] Intuitions may vary. If one thinks that a chance-related counterfactual such as 'If the coin had been tossed, it would have landed heads' is false and not merely unknown, one may lack the intuition about the corresponding instance of conditional excluded middle because one is inclined to reason that it has two false disjuncts.

instances of the schema ⌜$\phi > (\psi \vee \neg\psi)$⌝ which is valid according to any reasonable semantics for counterfactuals.

Lewis builds a case against conditional excluded middle by connecting the issue with the duality thesis:

> Another manifestation of Stalnaker's Assumption is that, except in the vacuous case, the difference between 'would' and 'might' counterfactuals is lost. [. . .] But surely English 'would' and 'might' counterfactuals do sometimes differ in truth value, and not only in the vacuous case. Stalnaker therefore cannot define the 'might' counterfactual as I have done. (Lewis 1973: 80)

No doubt that there is a difference between 'would'-counterfactuals and 'might'-counterfactuals. Lewis's claim that this difference would be lost if a Stalnaker semantics were adopted presupposes that a 'might'-counterfactual ⌜$\phi \diamond\!\!\rightarrow \psi$⌝ is—via the duality thesis—assigned the truth conditions of the corresponding 'would'-counterfactual ⌜$\neg(\phi > \neg\psi)$⌝. If 'might'-counterfactuals were given this semantics, then the relevant difference would be lost, for if the antecedent is not impossible, ⌜$\neg(\phi > \neg\psi)$⌝ is equivalent to ⌜$(\phi > \psi)$⌝ on a Stalnaker semantics and so, by duality, ⌜$\phi \diamond\!\!\rightarrow \psi$⌝ would be equivalent to ⌜$(\phi > \psi)$⌝.

There are a couple of things I would like to say in response. First of all, there are good reasons to reject the duality thesis.[14] But if duality fails, it is actually an advantage of the present account that it does not account for 'might'-counterfactuals in terms of 'would'-counterfactuals by deploying the duality thesis. Secondly, we have also seen how 'might'-counterfactuals may be assigned truth conditions without adopting the duality hypothesis. In fact, the proposed truth conditions agree (modulo the limit assumption—see next section) with the truth conditions Lewis assigns to them. It is only that the truth conditions for 'might'-counterfactuals are not defined in terms of 'would'-counterfactuals. But I see no prior reason why such an inter-definability should be a desideratum. As we would not expect that the truth conditions of simple 'might'-sentences can be defined in terms of simple 'will'-sentences, there is no prior reason to suppose that we can define the truth conditions of 'might'-counterfactuals in terms of 'would'-counterfactuals. Such a reason only emerges if we make the further assumption that 'would' is a necessity modal, but this is an assumption which can only be justified in the context

[14] Cf. Williams (2010), who defends conditional excluded middle along similar lines.

of a theory of counterfactuals—it is not a constraint we should impose prior to investigation.

7.4 The Limit Assumption

There is a background assumption of the present proposal which is sometimes considered to be controversial for a reason quite similar to why conditional excluded middle can be doubted. The semantics presupposes that there is always a set of relevant antecedent-worlds. The existence of such a set is what is sometimes questioned, or so it may seem.

As introduced by Lewis (1973: 19f.), the *limit assumption* says that to any antecedent which is not impossible, there is a set of closest antecedent-worlds. What the limit assumption thereby excludes is the possibility that there may be a world w and an antecedent ϕ, such that there are closer and closer ϕ-worlds without there ever being a set of closest ϕ-worlds (relative to w). Lewis finds the limit assumption doubtful:

> Unfortunately, we have no right to assume that there always are a smallest antecedent-permitting sphere and, within it, a set of closest antecedent-worlds. Suppose we entertain the counterfactual supposition that at this point there appears a line more than an inch long. (Actually it is just under an inch.) There are worlds with a line 2″ long; worlds presumably closer to ours with a line $1\frac{1}{2}''$; worlds presumably still closer to ours with a line $1\frac{1}{4}''$; worlds presumably still closer.... But how long is the line in the *closest* worlds with a line more than an inch long? (Lewis 1973: 20)

Assuming that the length of the line is the relevant dimension of comparison, Lewis gives here an example of an infinite sequence of antecedent-worlds which are closer and closer to the actual world. Of course, one may question this particular example (cf. the footnote in Lewis 1973: 20). But I see no reason to doubt that an example with the relevant structure can be given. For all we know, the present example might be just fine. Therefore, we should conclude that the limit assumption is false.

Does the failure of the limit assumption challenge the present account? The semantics presupposes that there is always a set of relevant antecedent-worlds. If the set of relevant antecedent-worlds is conceived of as the set of closest antecedent-worlds, then Lewis's example would show that there is not always a set of relevant antecedent-worlds. The present semantics would be based on a false background assumption. However, I do not want to make this identification: the set of relevant antecedent-worlds is not to be thought of as consisting exclusively of the closest worlds.[15]

To get an idea of what the conception of relevance which I have in mind amounts to, let us apply it to Lewis's case. In the right circumstances, I would assume that the set of relevant worlds may well contain for any lengths between 1 and 2 inches a world in which the line has that length. Of course, it will not contain for any length a world in which the line is of that length: no world in which the line covers the whole width of the page, say, may be relevant. There will be an upper limit. What the upper limit is will depend on the details of the example, for instance, on whether the author played with the thought of drawing a really long line or not, etc. Besides, we should reckon with the possibility that it will often be a vague matter what the upper limit is.

Let us look at a counterfactual which utilizes the supposition Lewis considers:

(23) If the line had been longer than one inch, it would have been $1\frac{1}{2}''$ long.

For Lewis, this counterfactual is false, I take it, and can, given appropriate circumstances, be known to be false (cf. Lewis 1973: 16).[16] On the present account, the counterfactual *might* be true. Worlds at which the line is $1\frac{1}{2}''$ long can be assumed to be among the relevant antecedent-worlds. But the set of relevant antecedent-worlds can also be assumed to contain worlds at which the line is shorter and worlds at which the line is longer. By the knowability constraint, we can therefore not know whether the line would have been $1\frac{1}{2}''$ long if it had been longer than one inch.

[15] Cf. Edgington (1995: 259). Bennett (2003: §70), following a suggestion by Stalnaker (1978), adopts the limit assumption but tries to construe it in a way different from Lewis by making closeness more coarse-grained and context-dependent so that in certain contexts different lengths of the line would not make for more or less closeness. If the context-dependence is subtracted from this proposal, then it seems to be quite similar to the present suggestion.

[16] Modifying the standard account accordingly comes at a cost, though. As Pollock (1976: 18–19) stresses, the inference from $\ulcorner\phi > \psi_i\urcorner$ and $\ulcorner\phi > \psi_j\urcorner$ to $\ulcorner\phi > (\psi_i \wedge \psi_j)\urcorner$ can no longer be iterated infinitely many times and thus shows a behaviour different from ordinary conjunction introduction. This point has recently been emphasized by Fine (2012a,b).

A consequence of this way of construing the set of relevant worlds is that the corresponding 'might'-counterfactual comes out true:

(24) If the line had been longer than one inch, it might have been $1\frac{1}{2}''$ long.

To my ears, this counterfactual sounds true. If so, this is some evidence for the package deal of combining a certain conception of relevance with a certain semantics for 'might'-counterfactuals.[17]

In sum, I suggest the following resolution of the issue concerning the limit assumption. Lewis is right in rejecting it. Sometimes there is no set of closest antecedent-worlds. But the set of relevant antecedent-worlds at work in a semantics for counterfactuals need not be assumed to consist solely of the closest antecedent-worlds. Rather, it may contain antecedent-worlds which are more or less close to the actual world. Then the truth conditions of counterfactuals can still be specified in terms of a set of relevant worlds.

[17] For Lewis, the 'might'-counterfactual comes out false, partly due to his semantics for 'would'-counterfactuals without the limit assumption, partly due to his account of 'might'-counterfactuals in terms of the duality thesis.

8

Triviality

For indicative conditionals, there is a certain class of formal arguments—the so-called *triviality results*—which make trouble for any truth-conditional semantics which is supposed to underpin something like Adams's thesis.[1] The triviality results received their name because they establish, in some way or other, that a given version of Adams's thesis can only be satisfied by a very small class of rather poor probability functions. Some, perhaps most of the authors which have provided a triviality result (e.g. Lewis 1976) interpret the triviality results as arguments against Adams's thesis.[2] If one instead holds on to Adams's thesis, then these results can be seen as arguments against truth conditions for indicative conditionals (cf. Edgington 1995: 271). Although the complexity of (some of) the triviality results makes it somewhat harder to evaluate their philosophical impact, they have proven to be very powerful and are often considered to be the biggest threat to anyone who tries to combine a version of Adams's thesis with truth conditions for indicative conditionals. Therefore, it is a natural question whether the evaluation constraint for counterfactuals as developed in chapter 3, which has a strong claim for being the counterfactual analogue of Adams's thesis, gives rise to similar triviality results for counterfactuals. As it turns out, this is indeed the case. Leitgeb (2012a) and Williams (2012) have provided triviality results for counterfactuals (see also fn. 21

[1] The original triviality results are due to Lewis (cf. Lewis 1976, 1986c). An overview of various results is given by Hájek and Hall (1994).

[2] In the context of the triviality results, Adams's thesis is usually taken to concern propositions, not sentences. Hence, on a sentential interpretation, the triviality results leave Adams's thesis unchallenged. However, if sound, they would show that one cannot reinterpret the probabilities of conditionals qua sentences as probabilities of propositions expressed by these sentences.

in the previous chapter).[3] This would challenge the present semantics, for it comes with the promise to underpin the evaluation constraint for more than a trivial class of probability functions. Before looking at these results, I will start by evaluating Lewis's discussion of imaging, for he proves not only an interesting tenability result for Stalnaker conditionals, but also shows that only Stalnaker conditionals can validate the relevant kind of evaluation constraint.

8.1 Lewis on Imaging

In his discussion of imaging, Lewis (1976) shows that the non-generalized imaging constraint can be satisfied by the corresponding Stalnaker truth conditions. Recall that the non-generalized version of imaging is defined by assuming uniqueness about the selected antecedent-worlds. So, for any antecedent A and any world w, the set of relevant antecedent-worlds $f_w(A)$ contains exactly one world, w_A. The Stalnaker truth conditions can then be defined by setting

(S) $(w \models A > B)$ iff $(w_A \models B)$,

i.e. a counterfactual is true at world w iff the consequent is true at the single most relevant antecedent-world w_A. The proposition expressed by the counterfactual is then the set of worlds $\{w : w_A \models B\}$. What is the probability of this set?

Suppose P is defined over a finite set of worlds W. Then

(1) $P(\{w : w_A \models B\}) = \sum_{w:w_A \models B} P(w) = \sum_{w' \in B} \sum_{w:w_A \models w'} P(w)$.

But the latter just is the non-generalized image $P^A(B)$ (cf. section 3.6). The probability of a Stalnaker conditional is the image of the consequent under the antecedent:

(2) $P^A(B) = P(A > B)$.

The reason is simple: in evaluating a Stalnaker conditional, we have to ask, for every epistemically possible world, whether the antecedent-world closest to it is a consequent-world. But this means that we could equally

[3] I have also learned through the references in Leitgeb (2012a) and Schwarz (ms) that Rachael Briggs has proven a triviality result for a closely related evaluation constraint in an unpublished manuscript.

well ask which of the worlds closest to some epistemically possible worlds are consequent-worlds. If we weight the worlds with probabilities, we see that in evaluating a Stalnaker conditional, we shift the probability of an epistemically possible world to the closest antecedent-worlds which are consequent-worlds.

So, Stalnaker truth conditions satisfy the corresponding non-generalized imaging constraint. This bears on possible triviality results in the following way: a triviality result which rules out the possibility of satisfying the non-generalized imaging constraint will show definitely too much.

Lewis (1976) has also proven an important converse of this: only Stalnaker conditionals satisfy suppositional constraints like the imaging constraint. The result is very general. We first need an abstract characterization of assigning a probability $P(B||A)$ to a proposition B under a supposition A. A natural idea is to generalize the basic properties of conditional probability. Lewis considers the following three constraints on suppositions (cf. also the discussion in Joyce 1999: ch. 6):

(C1) $P(A||A) = 1$.

(C2) If $P(A) = 1$, then $P(\bullet||A) = P$.

(C3) If $P(B||A) = 1$ and $P(A||B) = 1$, then $P(\bullet||A) = P(\bullet||B)$.

The first constraint requires that a supposition should be made certain under itself. The second requirement is that if the supposition is already certain, making it should not change anything. A supposition should be integrated by making minimal changes only—if none are needed, nothing gets changed. The idea of minimal revision is in the background of the third constraint as well. If integrating A at minimal costs makes B certain and integrating B at minimal costs makes A certain, then both suppositions require the same changes to integrate them (this is a probabilistic variant of the equivalence condition on relevance; cf. 3.1). It is easily verified that conditional probability and imaging, both the generalized and the non-generalized version, have these three features.

Now, given an arbitrary suppositional operation $||$, we may ask whether there is a conditional operator $\rightarrow$ which satisfies

(Sup) $P(B||A) = P(A \rightarrow B)$.

This is a generalization of Adams's thesis and the imaging constraint. For any kind of supposition, we may ask whether there is a conditional such that the probability of the conditional is the probability of the consequent under the supposition of the antecedent. Lewis shows that if a conditional satisfies this equation for all probability functions P and all sentences A and B, then it is a Stalnaker conditional, i.e. we can find a selection function such that the conditional is true at a world just in case the consequent is true at the single selected antecedent-world. In other worlds, only Stalnaker conditionals are capable of satisfying suppositional constraints like (Sup).

The idea is the following. Consider a probability function P_w which concentrates all probability on w. In other words, P is omniscient about w. Now, if we engage in a supposition from the perspective of P_w, what are the resulting probabilities, i.e. how does $P_w(\bullet||A)$ behave? The core observation Lewis makes is that the resulting probability function must also concentrate all probability on a single world. If it were otherwise, we would find a B such that $0 < P_w(B||A) < 1$. By (Sup), this would imply $0 < P_w(A \to B) < 1$, which cannot be, for P_w assigns only values of 0 and 1. As P_w is decided about all matters of fact including conditionals, it cannot assign any intermediate values. But then $P_w(\bullet||A)$ will also concentrate all probability on a single world w' so that we can set $w_A := w'$. This defines a Stalnaker conditional and given the properties of $||$, it is well behaved, i.e. it satisfies the constraints standardly imposed on a selection function. For instance, the first constraint, (C1), guarantees that the world w' is an A-world and (C2) secures that if A is already true at w, then $w' = w$.

In sum, a Stalnaker conditional underwrites imaging and only Stalnaker conditionals satisfy suppositional constraints like imaging. To some extent, this is reassuring, for the present semantics assigns counterfactuals Stalnaker type truth conditions by binding the truth-value of a counterfactual to what happens at a single world. On the other hand, it aims to validate the generalized version of imaging. For this reason, Lewis's observation does not yet provide a safe harbour for this account. As a matter of fact, we shall see shortly that the observation Lewis makes about decided probability functions like P_w can actually be levelled against any attempt to satisfy the generalized imaging constraint.

8.2 Leitgeb's Observation

Leitgeb (2012a) uses the the same observation Lewis makes to argue against the idea that the probability of a counterfactual could go with the corresponding conditional objective chance (see also the discussion in Schwarz ms). Although the proposal Leitgeb targets is not quite the evaluation constraint adopted here (he focusses on Skyrms's account), this difference proves inessential and I will discuss the objection as if it were directed against the generalized imaging account. Like Lewis, Leitgeb observes that for a probability function P_w centred on a single possible world, we have

(3) $P_w(A > B) = 1$ or $P_w(A > B) = 0$,

depending on whether or not the counterfactual is true at w.

Now suppose that we adopt the generalized imaging account. Then for some world w, the set of relevant A-worlds is not a singleton set and so there will be a possible consequent B such that the counterfactual chance of B under A relative to w will be an intermediate value, say 1/2. In this case, the image of B under A will be 1/2 relative to w and because P_w is not uncertain about which world is actual, the imaging constraint requires

(4) $P_w(A > B) = P^A(B) = 1/2$,

which contradicts our earlier conclusion that it must be either 1 or 0. This is a quick argument against any (bivalent) semantics which pretends to underpin the generalized imaging constraint. What can we say?

On an intuitive picture, one may feel tempted to deny the assumption that a probability function centred on a single world must assign maximal or minimal probability to any statement, including counterfactuals. If we think of P_w as comprising all the information which can possibly be gathered in w, it may still be that some propositions remain undecided. As we saw, the semantics in terms of the epsilon operator actually predicts that in a case in which the counterfactual chance assumes an intermediate value like 1/2, the counterfactual will be undecided in w. So, if P_w represents a possible state of knowledge in w, it will not decide the counterfactual.

However, thinking of P_w as a possible state of knowledge is probably not the intended interpretation relevant to a triviality result like Leitgeb's. Here P_w is simply constructed by conditionalizing a probability function on the singleton set containing w and it is assumed that a possible world is maximal in the sense that it decides all propositions, including counterfactuals. So, by construction, P_w assigns probability 1 or 0 to any counterfactual.

But if P_w is certain about the counterfactual, it evidently possesses inadmissible information. Direct information about the truth-value of a counterfactual is the clearest example of inadmissibility. In such a case, the evaluation constraint simply does not apply as we have detailed in our discussion in chapter 3 (Schwarz (ms) makes a similar point). The Principal Principle type formulation of it is explicitly cautious in this regard: the principle only applies to situations in which no inadmissible information is possessed. The imaging constraint does not come with such a proviso, but this is, as we argued, a reason to think that it may not hold in full generality. It was exactly because of this possible difference to the Principal Principle type formulation that we should feel committed to the imaging constraint only under the assumption that issues about admissibility do not play any role.

In sum, a credence function like P_w does not always represent a possible state of knowledge on the semantics I favour. More important in the context of the triviality result is, however, the observation that it contains inadmissible information by deciding the truth-value of the counterfactual. In such cases, the evaluation constraint on counterfactuals defended here is not applicable, for it is only intended for situations in which we do not possess information about the counterfactual which goes beyond the information about the counterfactual chances.

8.3 Williams's Extension of Lewis's Triviality Result

In a recent paper, Williams (2012) argues that Lewis's (1976) original triviality result can be extended to counterfactuals. His argument is based on Skyrms's constraint as discussed in chapter 2, but Williams reckons with the possibility that it may call for some restrictions. Lewis's original result aims at a reductio of the propositional version of Adams's thesis, namely that there is a connective $*$ such that for all propositions A and B, modelled as sets of possible worlds, and a class of probability functions P: $P(A * B) = P(B|A)$.

To begin with, let us note two basic facts about conditional probabilities. First, conditional probabilities, if defined, satisfy the following equation (where $\overline{B}$ is the complement of B, i.e. B's negation):

(5) $P(A) = P(A|B) \cdot P(B) + P(A|\overline{B}) \cdot P(\overline{B})$.

In words: the probability of a proposition A is, for any B, the probability of A given B weighted by the probability of B plus the probability of A given $\overline{B}$ weighted by the probability of $\overline{B}$. Formally, this follows from the definition of conditional probability as $\frac{P(A \cap B)}{P(B)}$. By multiplication with $P(B)$, this reduces the first summand to $P(A \cap B)$ and similarly the second one to $P(A \cap \overline{B})$. Hence, the result follows from $P(A) = P(A \cap B) + P(A \cap \overline{B})$: the probability of A can be packaged into the probability of $A \cap B$ and $A \cap \overline{B}$.

The second fact applies to two probability functions P_B and $P_{\overline{B}}$ which are defined as $P(\bullet|B)$ and $P(\bullet|\overline{B})$, that is, they come from P by conditionalizing on B and $\overline{B}$ respectively. Now note that $P_B(B|A) = 1$ because it equals $P(B|A \cap B)$; for analogous reasons, $P_{\overline{B}}(B|A) = 0$. This is already all Lewis needs.

To get his original result, we start by applying the first fact to the potential conditional proposition and get

(6) $P(A * B) = P(A * B|B) \cdot P(B) + P(A * B|\overline{B}) \cdot P(\overline{B})$.

By the definition of P_B and $P_{\overline{B}}$, this simplifies to

(7) $P(A * B) = P_B(A * B) \cdot P(B) + P_{\overline{B}}(A * B) \cdot P(\overline{B})$.

If we now apply the propositional version of Adams's thesis to P_B and $P_{\overline{B}}$, we find

(8) $P(A * B) = P_B(A|B) \cdot P(B) + P_{\overline{B}}(A|B) \cdot P(\overline{B})$.

Given the second fact about probabilities, the conditional probabilities are 1 and 0 respectively. We can therefore conclude that

(9) $P(A * B) = 1 \cdot P(B) + 0 \cdot \neg B = P(B)$.

The final conclusion says that the probability of the conditional must equal the probability of the consequent. However, this cannot generally be the case and contradicts Adams's thesis for many choices of the relevant probability function. We may note that Lewis's result only requires that if Adams's thesis holds for a probability function P, it also holds for probability functions which come from P by conditionalization.

Williams points out that the proof is formally valid and so does not need to presuppose anything about the kind of probability represented by P as long as the relevant kind obeys the standard laws of probability. For this reason, we may as well interpret the result as also showing that there is no conditional connective such that the *objective chance* of a conditional is always the corresponding conditional objective chance. Given that Skyrms's version of the evaluation constraint on counterfactuals operates with objective chances, this interpretative possibility is a potential source of trouble.

But in order to run Lewis's result on counterfactuals, Williams needs to bridge the gap between the subjective probabilities and the objective chances which are connected in the evaluation constraint, for the evaluation constraint does not say that the objective chance of a counterfactual is the corresponding conditional objective chance. It rather says that the subjective probability of a counterfactual should be so-and-so on the assumption that the corresponding objective chances are so-and-so.

To close this gap, Williams considers a scenario where all the relevant objective chances are known. We may assume that the the relevant subjective probability function P comes from an initial probability function by updating it (only) with the relevant chance information. This is helpful in two ways. First, it avoids the problem that the evaluation constraint on counterfactuals only makes a conditional claim, for if the chances are known, i.e. the subjective probability that they are so-and-so is 1, Skyrms's version forces them to actually match those chances. Hence, we get

(10) $P(A > B) = \text{Ch}(B|A)$.

Secondly, the assumption that all the chances are known helps extend Lewis's result also in another respect. If all the relevant chances are known, then in particular the chances of a counterfactual being true are known. This makes Lewis's original Principal Principle applicable, for given that P was only given information about chances, it is clear that it cannot contain any inadmissible information. Thus, we see that our credence in the counterfactual should also go with the objective chance of the counterfactual:

(11) $P(A > B) = \text{Ch}(A > B)$.

If we know that the objective chance of the counterfactual being true is $\text{Ch}(A > B)$, our credence in the counterfactual should be $\text{Ch}(A > B)$. But combining the two equations now gives us

(12) $\text{Ch}(A > B) = \text{Ch}(B|A)$.

This bridges the gap between subjective probabilities and objective chances. By the evaluation constraint, the subjective probability of a counterfactual should be the corresponding objective conditional chance. By Lewis's Principal Principle, the subjective probability of a counterfactual should also be the unconditional objective chance of the counterfactual. But then the conditional objective chance has to match the unconditional objective chance which instantiates a version of Adams's thesis with objective chances substituted for subjective probabilities. Given that the kind of probability does not matter for Lewis's proof, we can reduce a propositional interpretation of the resulting thesis to absurdity, or so Williams suggests.

There are a couple of issues with Williams's argument. The first question is perhaps whether the proof is formally valid. There is reason to doubt its validity having to do with the fact that in order to run Lewis's original triviality result on the chance function Ch, we do not only need the instance (12) = '$\mathrm{Ch}(A > B) = \mathrm{Ch}(B|A)$', but also two further instances which come from Ch by conditionalizing on B and $\overline{B}$ respectively:

(13) $\mathrm{Ch}_B(A > B) = \mathrm{Ch}_B(B|A)$

(14) $\mathrm{Ch}_{\overline{B}}(A > B) = \mathrm{Ch}_{\overline{B}}(B|A)$.

The question is whether these two further instances are also implied by the evaluation constraint on counterfactuals. If not, then the starting assumptions for Lewis's result are not met. Although the evaluation constraint would imply certain instances identifying the objective chance of a counterfactual with the corresponding conditional chance, it would not yield enough such instances for Lewis's result to kick in.[4]

Now, as far as I can see, these further instances are not implied by the evaluation constraint. To make the case more vivid, consider a simple coin toss counterfactual $\ulcorner A > B \urcorner$. The evaluation constraint on counterfactuals may then imply that the subjective probability of $\ulcorner A > B \urcorner$ should equal the objective chance of the coin coming up heads a certain while before it would have been tossed, say at time t. So, we should write more precisely

[4] Schwarz (ms) makes a similar point, but his argument is based on the observation that the relevant class of chance functions may not be closed under conditionalization, not that the relevant equations are not implied by the evaluation constraint on counterfactuals.

(15) $P(A > B) = \text{Ch}^t(A > B)$.

Suppose further we update the chance function with B, i.e. with the information that the coin came up heads. Is the equation

(16) $P(A > B) = \text{Ch}^t_B(A > B)$.

implied by the evaluation constraint? First of all, note that this equation is quite implausible. After updating, the chance of B, i.e. that the coin came up heads, is 1 and so our credence in the counterfactual would have to be 1. But nothing in the evaluation constraint suggests this: what is suggested is that the credences in counterfactuals should go with objective chances taken at certain points in time from which on the history of the relevant counterfactual worlds are allowed to depart from the actual history. It says nothing about how the credences in counterfactuals should relate to objective chances which are conditionalized on the information contained in the consequent of the counterfactual (or its negation). Hence, it seems we cannot simply run Lewis's triviality result on the instances of equations about objective chance, for the evaluation constraint does not imply enough of such instances. This does not mean that the equations it does imply might not be problematic, but Lewis's original result cannot be used to show this.

Another difficulty concerns the role of admissibility in the application of the evaluation constraint on counterfactuals. This problem is best seen when we consider Williams's argument as applied to the evaluation constraint defended in chapter 3, where it is given in the form of a Lewisian Principal Principle. I do think that the same problem already arises in the application of Skyrms's constraint, but it would be harder to make out. In any case, seeing how the argument fares under the constraint I ultimately favour is what presently matters most.

The evaluation constraint defended in chapter 3 is formulated in terms of conditional counterfactual chances and not in terms of ordinary physical chances. On the other hand, Lewis's Principal Principle is formulated in such terms. As a result, the two core assumptions about objective chance in Williams's argument will concern on the present interpretation, strictly speaking, two different kinds of probability: physical chances and counterfactual chances. To keep the two distinct, let us write, only for the purposes of the present discussion, 'PCh' for physical chances as they figure in Lewis's Principal Principle and 'CCh' for counterfactual chances as they figure in the evaluation constraint on counterfactuals (so far, we have occasionally used 'Ch' both for physical and for counterfactual chances).

Although we have seen that (conditional) physical chances and counterfactual chances do not always coincide, there are still many instances for which we can assume that they have the same value. A case in point is a coin toss counterfactual. We may assume that the conditional physical chance taken at the relevant point in time of the coin coming up heads conditional on being tossed is 1/2 just like the corresponding counterfactual chance. So, we can surely find a counterfactual $\ulcorner A > B \urcorner$ for which

(17) $\text{PCh}(B|A) = \text{CCh}(B|f_w(A)) = 1/2$.[5]

(Recall here that counterfactual chances are computed by conditionalizing on the set of relevant antecedent-worlds and not by conditionalizing on all antecedent-worlds.)

By identifying the two kinds of chance over a certain range of cases, Williams's argument regains its applicability. But something interesting happens to his assumption that all the relevant objective chances are known. In a first step, this assumption is used to apply to the evaluation constraint on counterfactuals in order to conclude that the subjective probability of the counterfactual matches the objective chances. Given that the evaluation constraint defended in chapter 3 is cast in terms of counterfactual chances, what has to be assumed is that the relevant counterfactual chances are known:

(18) $P(\text{CCh}(B|f_w(A)) = 1/2) = 1$.

This much is unproblematic: although we are sometimes uncertain about what the counterfactual chances are, we can in principle become certain about these matters. A simple coin toss counterfactual may well be a case in which we know the counterfactual chances. By the evaluation constraint on counterfactuals, it would follow that our credence in the counterfactual matches the corresponding counterfactual chance provided no

[5] It is perhaps worth noticing that on the non-generalized version of imaging, the counterfactual chances will always be 1 or 0, for $f_w(A)$ will be a singleton set. In this case, the value 1/2 is, of course, not possible nor is any other intermediate value. But if only the values 1 and 0 are assumed, a resulting equation of the form $\text{Ch}(A > B) = \text{Ch}(B|A)$ is unproblematic and does not run into triviality. So, the implicit assumption of intermediate values is why Williams's argument is compatible with Lewis's tenability result.

inadmissible information is possessed on top of that (this will become important later on):

(19) $P(A > B) = \mathrm{CCh}(B|f_w(A)) = 1/2.$

But now note that knowing the counterfactual chances does not make Lewis's Principal Principle applicable. In order to apply it, we also need to know the relevant physical chances, in particular the unconditional physical chance of the counterfactual being true:

(20) $P(\mathrm{PCh}(A > B) = x) = 1$, for some x.

With Lewis's Principal Principal, we can then conclude that

(21) $P(A > B) = \mathrm{PCh}(A > B).$

Our credence in the counterfactual matches its objective chance. Given that it also matches the corresponding counterfactual chance and that we have chosen the example so that the latter does not differer from the conditional physical chance in this particular case, we can still derive the problematic equation

(22) $\mathrm{PCh}(A > B) = \mathrm{PCh}(B|A).$

The physical chance of the counterfactual would equal the corresponding conditional physical chance.

Although we have already seen that this equation does not lead to triviality via Lewis's original result, it could still be a potential source of trouble. For this reason, it is worth seeing whether we can avoid this conclusion. It has been derived in much the same way as Williams suggests in his argument, but we had to make an additional assumption about what the respective agent is supposed to know. Not only did we have to assume that the agent knows the conditional counterfactual chances, the agent was also supposed to know the unconditional physical chance of the counterfactual being true (see (18) and (20)). The former assumption is uncontroversial: conditional counterfactual chances are frequently known. The latter assumption, on the other hand, is very strong: knowing the unconditional physical chance of a counterfactual being true seems hard. When the truth value of a counterfactual is still undecided, how can we know the unconditional chance of a counterfactual being true? It does not seem that we have any direct epistemic access to facts of this kind. Of course, we do have access to conditional physical chance.

If there were a valid bridge principle linking conditional physical chances of counterfactuals to unconditional ones, this would explain how such chances could be known. But the assumption of such a bridge principle is what Williams's argument is directed against and what we are presently looking to avoid. So, it seems like an open question how the assumption that the unconditional physical chances are known could be justified.

But perhaps these kinds of chances are only hard but not impossible to know. So, let us grant for the sake of argument that they can be known. This leads to what I take to be the most serious problem with this assumption. We have already mentioned in chapter 3 that information about unconditional physical chances is likely to be inadmissible in the evaluation constraint on counterfactuals. Lewis's original Principal Principle and the structurally similar formulation of the evaluation constraint on counterfactuals come with different conceptions of admissibility. We have put the difference in terms of *o*-admissibility versus *c*-admissibility. For instance, although information about the future is inadmissible in the original principle (if it bears on the proposition in question), it is admissible in the evaluation of counterfactuals. On the other hand, information about unconditional chances, which is clearly admissible in the original principle, is not generally admissible in the evaluation of counterfactuals. This is most easily seen for cases in which the unconditional chance of a counterfactual is 1 or 0 at some late point t in time simply because the truth-value of the counterfactual is settled at t. But these may not be the only cases: the unconditional chances of counterfactuals may bear more directly on the truth-value of a counterfactual than the corresponding conditional counterfactual chances. For this reason, it is possible to reject the assumption that the unconditional physical chances are known as inadmissible in the application of the evaluation constraint on counterfactuals. In practice, this issue is unlikely to ever come up, for, as argued above, we will hardly, if ever, have any direct information about the unconditional chances of counterfactuals.

8.4 Conclusion

In closing, let me address a potential puzzlement one could have with respect to the position I have been defending in response to Williams's argument. It might seem that equating the unconditional physical chance of a counterfactual with the corresponding conditional chance is quite congenial to the overall spirit of the picture I have been pushing for. After all, this assumption is not so different from the suggestion that the subjective probabilities of counterfactuals should equal, on ideal reflection, certain conditional chances. In both cases, conditional probabilities are linked with unconditional ones. There are, of course, some differences. The subjective probabilities of counterfactuals do not quite go with the corresponding chances conditional on the antecedent, but rather with the chances conditional on the relevant antecedent-worlds. Moreover, they are not supposed to always match them, but rather to estimate them. All this led to describing the subjective probabilities of counterfactuals in terms of the probability of the consequent under the image of the antecedent.

On the theory I am envisaging, there is a further reason to be sceptical about a direct link between physical chances of counterfactuals and conditional physical chances. The explanation of the subjective probabilities of counterfactuals rests on the specific epistemic profile induced by the epsilon operator. As was pointed out in section 6.2.2, it is to be expected that the epistemic profile of counterfactuals does not go hand in hand with their modal profile. Whereas subjective probabilities target the epistemic profile of sentences, objective chances are sensitive to their modal profile. While the subjective probability that Hesperus is Phosphorus can rationally assume any value between 0 and 1, the objective chance of this proposition is always 1 due to its metaphysical necessity. So, on the present theory, the link between the subjective probabilities of counterfactuals and conditional chances essentially depends on the specific epistemic profile counterfactuals are assumed to have. There is no direct way of extending this link to the level of objective chances, for the latter are only sensitive to the modal profile of the relevant sentence or proposition.

To sum up, I have pursued two lines of defence against the threat of triviality. The first one is to see whether the probability functions constructed in the course of a triviality result contains any inadmissible information. It is at this point where it is crucial to keep in mind that just like the original Principal Principle, the evaluation constraint on counterfactuals is only applicable to probability functions not containing inadmissible information. So, in the context of possible triviality results, the focus should be on the Principal Principle type formulation of the evaluation constraint and not on the one in terms of imaging, for the imaging constraint is overly general in this regard. A second line of defence concerns the possibility of identifying under certain conditions subjective probabilities and more objective kinds of probability in order to make the original results for indicative conditionals applicable to counterfactuals. Here it is important to be aware that even if some such identifications are possible, it does not follow that we get a full-blown version of Adams's thesis with objective chances substituted for subjective probabilities. The evaluation constraint on counterfactuals does not imply an objectified version of Adams's thesis.

9

Concluding Remarks

In situations of uncertainty, counterfactuals are evaluated in a puzzling way. Our uncertain judgements about counterfactuals deviate dramatically from what standard semantics would predict. Finding a theory of counterfactuals which can explain counterfactual reasoning in situations of uncertainty was the main goal of this study.

The proposed semantics comes with a somewhat refined picture of counterfactuals. Whereas counterfactuals are in standard semantics conceived of as expressing a certain kind of *restricted necessity*—truth of the consequent in *all* relevant antecedent-worlds—the present theory describes counterfactuals as being of intermediate strength: weaker than counterfactual necessity but stronger than counterfactual possibility. In this regard, counterfactuals resemble future contingents. The truth-value of a future contingent is eventually settled by how the world happens to unfold. Something similar is not possible in the counterfactual case: we cannot wait and see how a counterfactual future evolves. This may be the source of why counterfactuals involve an element of arbitrariness: which counterfactual future fixes the truth-value of a counterfactual is to some extent an arbitrary affair.

Quite clearly, this can only be a first step in resolving the problems surrounding counterfactuals and probability. There are quite a few issues left open. These concern the prospects of explaining the relevant data pragmatically or error-theoretically, or through a combination of two such approaches. It would also be desirable to go back to the data about indicative conditionals and see whether a similar explanation in terms of the epsilon-operator can be given there. By way of conclusion, let me briefly comment on these issues.

9.1 Error Theories

The problems surrounding conditionals and probability are multifarious. More often than not the relevant evidence seems to pull in different directions. In the light of this, one may expect the best overall theory to be best only by a small margin, emerging as the winner only after a comprehensive cost-benefit analysis. If one thinks that any account connecting conditionals with conditional probabilities will have some costs, one may want to see whether a positive account of the data does in fact fare better than an error theory. According to an error theory, the data we have considered to motivate the present theory should not be taken to be genuine. Rather, they would be described as reflecting a certain kind of error we are prone to making in our use of conditionals in contexts of uncertainty.

An error theory is probably best pictured as advocating a simple and straightforward semantics for counterfactuals, standard semantics or a version of a strict analysis, say. The data which suggest that a different semantics is called for would then have to be explained away as stemming from a certain kind of error. For instance, one might hold that we are simply unjustified in being fairly confident that we would have lost if we had bought a lottery ticket. Really, we should not invest any credence in this counterfactual, for it is knowably false, as predicted, for instance, by the standard account. To a large extent, the way we think about counterfactuals would be fundamentally mistaken. On the linguistic level, the interaction of counterfactuals with epistemic phrases would be seen as simply reflecting this mistake. Accordingly, there would be no need to explain how a sentence like 'It is likely that if I had bought a lottery ticket, I would have lost' can be true (or acceptable) in a given situation because it would, despite appearances, not be true (or acceptable) at all.

As I see it, an error theory of the envisaged kind would be particularly costly, for the kind of error we are supposed to make would be fairly global, quite systematic and to a large extent irreversible. We would be mistaken about almost all types of counterfactuals and in evaluating counterfactuals we would be following a principle, namely the evaluation constraint, for which there is, in fact, no rationale. Given the pervasiveness of the postulated error, a theory of this kind strikes me as a last resort. Now, this is not to say that an error theory must be false. After all, errors of the envisaged kind can, and perhaps have been made. But as long as we can give a positive account which is not too costly—and both the modifier view sketched in chapter 5 and the theory I ultimately favour seem to satisfy this condition—a brute error theory seems to be the least attractive option.

To be fair, we should note that there are ways to soften the consequences of an error theory. For a start, one may offer truths in the neighbourhood of the predicted falsehoods. Although most of the sentences which constitute evidence for the evaluation constraint would not be true or acceptable, certain variants of them may be true or acceptable. Ascriptions of subjective probability to the whole counterfactual may be substituted by ascriptions of objective chance which attach to the consequent of the counterfactual only. Such a strategy seems to be employed by Bennett (2003: ch. 6) (see, in particular, Bennett 2003: 251) and is also considered by Hájek (2007) in response to a slightly different but related problem (see the brief discussion in section 7.2). To illustrate the approach, consider one of the sentences we considered:

(1) It is likely that if I had bought a lottery ticket, I would have lost.

Although this sentence comes out false on an error theory, the suggestion would be that a close variant of the sentence would still be true:

(2) If I had bought a lottery ticket, there would have been a high chance that I would lose.

On this strategy, a wide scope epistemic expression would be replaced by a narrow scope objective expression.

The idea that sentences of the latter kind are true or acceptable when sentences of the former kind merely appear to be acceptable seems to work for some cases, but it will not always work. A first type of counterexample are cases in which the epistemic expression reflects an estimate of the corresponding counterfactual chances (this was mentioned already in the introduction). It might be an epistemic possibility that the lottery was rigged in my favour which would have made we a winner no matter what. If this possibility is sufficiently unlikely, we would still think it likely that we would have lost if we had bought a ticket. But it is not clear that the corresponding counterfactual with the objective chance ascription in the consequent would be acceptable for us when construed along the lines of standard semantics. There will then be an epistemic possibility according to which our ticket would have won no matter what. Hence, a counterfactual saying that there would have been a high chance of losing does not seem to be fully acceptable.

Another type of counterexample comes from cases where the chances do not seem to be the same across all relevant worlds (Hájek considers similar cases). Suppose me buying a lottery ticket sometimes but not very often causes an assistant in the company to rig the lottery in my favour. Then it seems that in some relevant worlds the lottery will be rigged, so it will not be true that the chance of my ticket losing will be high in all relevant worlds. On standard semantics, the narrow scope counterfactual will thus be false. On the other hand, given that the assistant does not help me most of the time and given that without any help it is vastly unlikely that I win, I should still think it likely that if I had bought a ticket, I would have lost.

It seems, then, that the strategy of offering acceptable sentences in the neighbourhood of the erroneous counterfactuals can at best be a partial success. Of course, one would like to check whether there are better alternatives to the strategy considered here, but I am not aware of any promising candidates (see also section 1.8 of the introduction). It would therefore seem that an error theorist should look for further ways of diminishing the costs of her theory.

An error theorist can give a fairly good explanation of how an error in evaluating counterfactuals could come about. For example, Leitgeb (2012a,b) (see also Leitgeb 2007) suggests that the troublesome credences may well not be credences that the proposed truth conditions obtain but rather—following Adams (1975, 1976) and Edgington (1995, 2008)—credences arrived at under the counterfactual supposition of the antecedent. According to Leitgeb, there are two ways of engaging epistemically with a counterfactual: (i) estimating how likely the truth conditions are satisfied and (ii) estimating how likely the consequent is satisfied under the counterfactual supposition of the antecedent. In a context of utterance, the second kind of credence might be pragmatically conveyed despite the fact that the first kind will always be literally communicated.

The challenge for such a dual account would be to uncover cases where our epistemic attitudes do indeed reflect what is predicted by (i) on the proposed truth conditions. Is there a sense in which we deem it to be certainly false that if we had bought a lottery ticket, we would have lost? As far as I can see, evidence that counterfactuals play a two-fold role in how they interact with epistemic modals and epistemic attitude verbs is hard to find, but I do take this to be an important question for future research. Another question would be by which process the second kind of credences, the ones which are captured by the evaluation constraint, would be pragmatically conveyed. I briefly comment on this in the next section, for at this point an error theory would actually join forces with a pragmatic explanation of the relevant data.

9.2 Pragmatic Explanations

Could the data be explained pragmatically? Some such story should accompany a possible error theory, but it also seems an interesting option in its own right, the more so the more systematic a story one could tell. As briefly mentioned in the introductory chapter, there have been attempts to explain the interaction of indicative conditionals and conditional probabilities in pragmatic terms. Perhaps the most influential account of this kind was Jackson's (1979, 1987) idea that the assertion of an indicative conditional conventionally implies that the corresponding conditional probability is high (see also the postscript to Lewis 1976 in Lewis 1986b[1]). Something similar could be tried for counterfactuals. Simply assume that the assertion of a counterfactual conventionally implies that our estimate of the corresponding counterfactual chances is high. We should observe, however, that this explains at best a small portion of the relevant data. It could explain why we might still find a counterfactual like 'If I had bought a lottery ticket, I would have lost' assertable (if we do find it assertable—I, for one, do not think that it is assertable; see section 6.3). But as it stands this approach leaves the interaction of conditionals with epistemic phrases completely unexplained. Why is the qualified sentence 'It is likely that if I had bought a lottery ticket, I would have lost' acceptable? Relatedly, how are the puzzling credences in counterfactuals to be explained? Something more, or perhaps something else entirely would be needed to give a pragmatic explanation of the problematic data.

A more promising approach would be, I think, to explore the possibility of a pragmatic take on the modifier view. On the modifier view, counterfactual 'if'-clauses are capable of modifying the domain of an epistemic expression. A pragmatic interpretation of this approach could

[1] It seems an open question to me what connection, if any, Lewis saw between Jackson's approach and the restrictor analysis he proposed elsewhere.

be derived in analogy to pragmatic constraints on quantifier domains. Context often imposes restrictions on the domain of a quantifier and depending on one's theory of domain restriction, some or even most restricted interpretations are derived pragmatically. In a similar vein, one might hold that counterfactual 'if'-clauses induce a pragmatic constraint on the interpretation of epistemic domains by raising to salience a modified interpretation. On such a picture, the problematic data would have a pragmatic explanation which stays very close to the modifier view as described in chapter 5.

Of course, a pragmatic variant of the modifier view can be expected to inherit the problem of embeddings the modifier view seems to have. But given that a pragmatic approach does not necessarily have to aim at a strict rule through which the interpretations of the problematic sentences can be derived, it is possible that a more lenient view on this problem might be defendable. Although one would have to see what the details will look like, a pragmatic approach of this kind seems to be a good fallback position if neither the positive theory defended here nor the semantic construal of the modifier view can be brought to work.

9.3 Unifying the Theory?

Perhaps the most important open problem concerns the possibility of a unified theory of conditionals. I have not tried to relate the present theory of counterfactuals to indicative conditionals. But given that the data concerning the two kinds of conditionals are quite similar, one may reasonably hope for a similar explanation in both cases.

It is not very hard to give an outline of what a corresponding theory for indicative conditionals might look like. Assume that the domain of the epsilon-operator are the contextually relevant epistemic possibilities. Assume further that there is a subjective probability distribution P over them. Now suppose that the epsilon-operator selects one of the epistemic antecedent-worlds with the conditional probability induced by P.

Formally, the truth conditions would look very much like the truth conditions for counterfactuals, with the epistemically possible antecedent-worlds in the indicative case playing the role of the relevant antecedent-worlds in the counterfactual case (cf. especially the explication of truth conditions at the end of section 6.2.1). They could be displayed roughly like this, with C_w being a set of salient epistemically possible worlds and P_w being a probability distribution over C_w (relative to a world w):

Possible Truth Conditions for Indicatives. Let w be any world and $\ulcorner\phi \rightarrow \psi\urcorner$ an indicative conditional. Then $w \models \ulcorner\phi \rightarrow \psi\urcorner$ iff

$$(\exists w' \in (C_w \cap [\![\phi]\!])) \supset (\epsilon_{P_w} w'(w' \in (C_w \cap [\![\phi]\!])) \models \psi).$$

According to these truth conditions, an indicative conditional is vacuously true iff its antecedent is epistemically impossible. Otherwise, it is true iff the consequent is true at an arbitarily selected epistemically possible antecedent-world.

The difference between indicative conditionals and counterfactuals would be mainly a difference in the kind of modality they are associated with: epistemic in the indicative case, metaphysical in the counterfactual case (cf. Weatherson 2001). A further difference would be a build-in context-dependence of indicative conditionals. As in the case of the restrictor view, one would not necessarily have to assume that indicative conditionals are always sensitive to the domain of overt epistemic modals. One could even allow for contexts C containing only the world of evaluation, so that indicative conditionals could be read as material conditionals.

A number of issues would have to be addressed. Can one justify a version of Adams's thesis on the basis of these truth conditions (or slight variants thereof)? *Prima facie* at least, there is hope for a positive answer. If P corresponds to our subjective probabilities, then the probability that a consequent-world gets selected seems to be the relevant conditional probability.[2] Which role exactly should be played by context, how is it determined and what constrains it? How should the logic of indicative conditionals be approached? Here, the discussion of the logic of counterfactuals could provide some guidance. Moreover, there are a great many further puzzles and problems about indicative conditionals which would have to be taken into view. A whole lot more would have to be said.

Just as in the counterfactual case—probably even more so—, a prime rival of such an approach would be an account along the lines of the restrictor view. An important question would therefore be whether the

2 Cf. the recent work in Bacon (2015).

problem of embeddings as discussed in section 5.7 can be reproduced for indicative conditionals. *Prima facie*, this seems to be the case. Consider:

(3) It is likely that if Jacky does not play this weekend, she'll get angry.

(4) Therefore, it is likely that there is someone who will get angry if she does not play this weekend.

The inference seems valid. There is an inconsistency in saying 'It is likely that if Jacky does not play this weekend, she'll get angry, but it is not likely that there is someone who will get angry if she does not play this weekend'. To make the problem more vivid, suppose that Jacky tends to get angry when she is not on the team, but occasionally accepts sitting on the bench. Moreover, no other player is determined to get angry on that condition, although some very well might. Given such a background, we cannot reconstruct the embedded conditional as restricting a covert necessity operator, neither in the premise nor in the conclusion (though see Rothschild 2013). On the other hand, the 'if'-clause in the conclusion cannot be assumed to restrict the probability operator due to the interfering quantifier. As in the counterfactual case, the most natural reaction would be to account for the conditional construction by finding a (covert) operator so that the probability of the conditional corresponds to the conditional probability. The epsilon-operator promises to do just that.[3]

One will have to control for the possibility that the problematic inference cannot be explained in some other way. Perhaps the nominal quantifier should be reconstructed as scoping over the epistemic phrase and out of the 'that'-clause (*prima facie* this strikes me as implausible; it also runs against epistemic modals tending to take widest scope). Perhaps the conditional in the conclusion is interpreted as the material conditional? That would make the embedded quantified conditional certain (which it does not seem to be), for it would already be verified by one player playing, which is something bound to happen. Perhaps the embedded

[3] Having a conditional proposition available for anaphoric reference would help with 'that's likely'-data as discussed in von Fintel (2003, 2007) (but see Kratzer 2012: 108). There is also some independent evidence for Stalnaker-type truth conditions stemming from embeddings of conditionals under quantifiers. See Higginbotham (1986, 2003) and Klinedinst (2011).

conditional is given a more objective interpretation (would this still validate the inference?).[4]

As far as I am aware, the transition problem for indicative conditionals does not extend to all constructions for which the restrictor view has a plausible story. For instance, it does not seem to extend to adverbs of quantification. Therefore, the proposal would not be to substitute the envisaged truth conditions for the restrictor account across the board, unless one finds alternative explanations for the remaining cases which suggest a restrictor story. In other words, an epsilon-based semantics for conditionals need not be seen as opposing the restrictor view of conditionals, for the epsilon-operator can naturally be construed as as another option for the implicit operator the restrictor view postulates for unembedded conditionals.

More generally speaking, I am open to the possibility that the aim of giving a unified account may lead to a reassessment of the weight of certain arguments I have given. If, for instance, it should turn out that the present account cannot be plausibly extended to indicative conditionals and the restrictor view proved to be the only viable explanation of how indicative conditionals embed under quantifiers and operators, it would be tempting to instead adopt the modifier view of counterfactuals—which squares well with a restrictor view on indicatives—and somehow live with the costs.

9.4 Conclusion

Uncertain counterfactuals constitute, as I see it, a deep challenge to our understanding of how their semantics works. Both the credences we are inclined to assign to counterfactuals and the epistemic phrases we attach to them form a very puzzling pattern. As we have seen, the problem might even run so deep that the very possibility of associating counterfactuals with truth conditions is called into question. I think the standard account alone, without any extra pragmatic or semantic elements added to it, cannot meet the challenge. In my eyes, the modifier view, construed either semantically or pragmatically, stands a good chance of providing a satisfying account. An even better way, I believe, is to

[4] I received valuable input on these issues from the participants of a workshop on conditionals in Hamburg 2014.

construe 'would'-counterfactuals in close analogy to 'will'-sentences. Just like a 'will'-sentence typically concerns a single and as yet unknown or even undetermined future, 'would'-sentences are about single worlds to which we do not always have full epistemic access. If a counterfactual (or, more generally, any conditional) has a false antecedent, the counterfactual world associated with it is not fully determined. This is, I suggest, the reason why the semantics of counterfactuals should contain an arbitrary element: the single counterfactual world we seem to think and talk about can only be fixed arbitrarily.

Bibliography

Ackermann, W. 1924: 'Begründung des "tertium non datur" mittels der Hilbertschen Theorie der Widerspruchsfreiheit'. *Mathematische Annalen* 93, pp. 1–36.

Adams, E. W. 1965: 'A Logic of Conditionals'. *Inquiry* 8, pp. 166–97.

Adams, E. W. 1970: 'Subjunctive and Indicative Conditionals'. *Foundations of Language* 6, pp. 89–94.

Adams, E. W. 1975: *The Logic of Conditionals*. Dordrecht: Reidel.

Adams, E. W. 1976: 'Prior Probabilities and Counterfactual Conditionals'. In Harper and Hooker (1976), pp. 1–21.

Adams, E. W. 1996: 'Four Probability-Preserving Properties of Inferences'. *Journal of Philosophical Logic* 25, pp. 1–24.

Ahmed, A. 2011: 'Walters on Conjunction Conditionalization'. *Proceedings of the Aristotelian Society* 111, pp. 115–22.

Anderson, A. R. 1951: 'A Note on Subjunctive and Counterfactual Conditionals'. *Analysis* 12, pp. 35–8.

Asser, G. 1957: 'Theorie der logischen Auswahlfunktionen'. *Zeitschrift für mathematische Logik und Grundlagen der Mathematik* 3, pp. 30–68.

Bacon, A. 2015: 'Stalnaker's Thesis in Context'. *The Review of Symbolic Logic* 8, pp. 131–63.

Bacon, A. ms: 'In Defence of a Naïve Conditional Epistemology'.

Barnett, D. 2006: 'Zif is If'. *Mind* 115, pp. 519–66.

Barnett, D. 2010: 'Zif Would Have Been If: A Suppositional View of Counterfactuals'. *Noûs* 44, pp. 269–304.

Barwise, J. and R. Cooper 1981: 'Generalized Quantifiers and Natural Language'. *Linguistics and Philosophy* 4, pp. 150–219.

Belnap, N. D. 1970: 'Conditional Assertion and Restricted Quantification'. *Noûs* 4, pp. 1–12.

Bennett, J. 1974: 'Counterfactuals and Possible Worlds'. *Canandian Journal of Philsophy* 4, pp. 381–402.

Bennett, J. 1984: 'Counterfactuals and Temporal Direction'. *Philosophical Review* 93, pp. 57–91.

Bennett, J. 2003: *A Philosophical Guide to Conditionals*. Oxford: Oxford University Press.

Bhatt, R. and R. Pancheva 2006: 'Conditionals'. In Evesaest, M. and H. von Riemsdijk (eds.): *The Blackwell Companion to Syntax*, vol. 1, Blackwell Publishing, pp. 638–87.

Bigelow, J. C. 1976: 'Possible worlds foundations for probability'. *Journal of Philosophical Logic* 5, pp. 299–320.

Breckenridge, W. and O. Magidor 2012: 'Arbitrary Reference'. *Philosophical Studies* 158, pp. 377–400.

Brogaard, B. and J. Salerno 2013: 'Remarks on counterpossibles'. *Synthese* 190, pp. 639–60.

Chalmers, D. J. 2006: 'The Foundations of Two-Dimensional Semantics'. In García-Carpintero, M. and J. Marcià (eds.): *Two-Dimensional Semantics*, Oxford: Oxford University Press, pp. 55–140.

Chisholm, R. M. 1946: 'The Contrary-to-Fact Conditional'. *Mind* 55, pp. 289–307.

Cross, C. B. 2006: 'Conditional Logic and the Significance of Tooley's Example'. *Analysis* 4, pp. 325–35.

Cross, C. B. 2008: 'Antecedent-Relative Comparative World Similarity'. *Journal of Philosophical Logic* 37, pp. 101–20.

Cross, C. B. 2009: 'Conditional Excluded Middle'. *Erkenntnis* 70, pp. 173–88.

DeRose, K. 1999: 'Can It Be That It Would Have Been Even Though It Might Not Have Been?' *Philosophical Perspectives* 13, pp. 385–413.

Dudman, V. H. 1983: 'Tense and Time in English Verb Clusters of the Primary Pattern'. *Australian Journal of Linguistic* 3, pp. 25–44.

Dudman, V. H. 1984: 'Conditional Interpretations of *If*-Sentences'. *Australian Journal of Linguistics* 4, pp. 143–204.

Dudman, V. H. 1988: 'Indicative and Subjunctive'. *Analysis* 48, pp. 113–22.

Dudman, V. H. 1991: 'Interpretations of 'If'-Sentences'. In Jackson (1991), pp. 202–32.

Dummett, M. 1973, 1981: *Frege: The Philosophy of Language*. London: Duckworth. Page references to the 2nd edition.

Eagle, A. ms: ''Might' Counterfactuals'.

Edgington, D. 1986: 'Do Conditionals Have Truth Conditions?' *Critica* 18, pp. 3–30.

Edgington, D. 1995: 'On Conditionals'. *Mind* 104, pp. 235–329.

Edgington, D. 2000: 'General Conditional Statements: A Response To Kölbel'. *Mind* 109, pp. 109–16.

Edgington, D. 2004a: 'Counterfactuals and the Benefit of Hindsight'. In Dowe, P. and P. Noordhof (eds.): *Cause and Chance—Causation in an Indeterministic World*, London: Routledge, pp. 12–27.

Edgington, D. 2004b: 'Two Kinds of Possibility'. *Proceedings of the Aristotelian Society. Supplementary Volume* 78, pp. 1–22.

Edgington, D. 2008: 'Counterfactuals'. *Proceedings of the Aristotelian Society* 108, pp. 1–21.

Eells, E. and B. Skyrms (eds.) 1994: *Probability and Conditionals*. Cambridge: Cambridge University Press.

Égré, P. and M. Cosiz 2011: 'If-Clauses and Probability Operators'. *Topoi* 30, pp. 17–29.

Etlin, D. 2008: *Desire, Belief, and Conditional Belief*. Ph.D. thesis, MIT.

Field, H. 1986: 'Critical Notice: Robert Stalnaker's *Inquiry*'. *Philosophy of Science* 53, pp. 425–48.

Fine, K. 1975: 'Critical Notice'. *Mind* 84, pp. 451–8.

Fine, K. 2012a: 'Counterfactuals Without Possible Worlds'. *Journal of Philosophy* 109, pp. 221–46.

Fine, K. 2012b: 'A difficulty for the possible worlds analysis of counterfactuals'. *Synthese* 189, pp. 29–57.

Gärdenfors, P. 1988: *Knowledge in Flux*. Cambridge, MA: MIT Press.

Geurts, B. 2004: 'On an Ambiguity in Quantified Conditionals', ms. Available online at https://sites.google.com/site/brtgrts/not-published. Last accessed: 13 June 2013.

Gibbard, A. 1981: 'Two Recent Theories of Conditionals'. In Harper et al. (1981), pp. 211–48.

Gillies, A. S. 2004: 'Epistemic Conditionals and Conditional Epistemics'. *Noûs* 38, pp. 585–616.

Gillies, A. S. 2009: 'On Truth-Conditions for *If* (but Not Quite Only *If*)'. *Philosophical Review* 118, pp. 325–49.

Glynn, L. 2010: 'Deterministic Chance'. *The British Journal for the Philosophy of Science* 61, pp. 51–80.

Greenough, P. and D. Pritchard (eds.) 2009: *Williamson on Knowledge*. Oxford: Oxford University Press.

Hájek, A. 2003: 'What Conditional Probability Could Not Be'. *Synthese* 137, pp. 273–323.

Hájek, A. 2007: *Most Counterfactuals are False*. Ms, in progress.

Hájek, A. 2014: 'Probabilities of Counterfactuals and Counterfactual Probabilities'. *Journal of Applied Logic* 12, pp. 235–52.

Hájek, A. and N. Hall 1994: 'The Hypothesis of the Conditional Construal of Conditional Probability'. In Eells and Skyrms (1994), pp. 75–111.

Hall, N. 1994: 'Correcting the Guide to Objective Chance'. *Mind* 103, pp. 505–17.

Harper, W. and C. A. Hooker (eds.) 1976: *Foundations of Probability Theory, Statistical Inference, and Statistical Theories of Science*, vol. 1. Dordrecht: Reidel.

Harper, W. L., R. C. Stalnaker, and C. T. Pearce (eds.) 1981: *Ifs*. Dordrecht: Reidel.

Haverkamp, N. and M. Schulz 2011: 'A Note on Comparative Probability'. *Erkenntnis*, pp. 395–402.

Hawthorne, J. 2005: 'Chance and Counterfactuals'. *Philosophy and Phenomenological Research* LXX, pp. 396–405.

Hawthorne, J. and M. Lasonen-Aarnio 2009: 'Knowledge and Objective Chance'. In Greenough and Pritchard (2009), pp. 92–108.

Higginbotham, J. 1986: 'Linguistic Theory and Davidson's Program in Semantics'. In LePore, E. (ed.): *Truth and Interpretation. Perspectives on the Philosophy of Donald Davidson*, Oxford: Basil Blackwell, pp. 29–48.

Higginbotham, J. 2003: 'Conditionals and Compositionality'. *Philosophical Perspectives* 17, pp. 181–94.

Hilbert, D. 1922: 'Neubegründung der Mathematik: Erste Mitteilung'. *Abhandlungen aus dem mathematischen Seminar der Hamburgischen Universität* 1, pp. 157–77, reprinted (with notes by Bernays) in Hilbert 1935: 157–77.

Hilbert, D. 1923: 'Die logischen Grundlagen der Mathematik'. *Mathematische Annalen* 88, pp. 151–65, reprinted in Hilbert 1935: 178–91.

Hilbert, D. 1935: *Gesammelte Abhandlungen*, vol. 3. Berlin: Springer.

Iatridou, S. 2000: 'The Grammatical Ingredients of Counterfactuality'. *Linguistic Inquiry* 31, pp. 231–70.

Jackson, F. 1979: 'On Assertion and Indicative Conditionals'. *Philosophical Review* 88, pp. 565–89, reprinted in Jackson 1991: ch. 6.

Jackson, F. 1987: *Conditionals*. Oxford: Basil Blackwell.

Jackson, F. (ed.) 1991: *Conditionals*. Oxford: Clarendon Press.

Jeffrey, R. 1965: *The Logic of Decision*. Chicago: University of Chicago Press.

Jeffrey, R. 1991: 'Matter of Fact Conditionals'. *Proceedings of the Aristotelian Society Supplementary Volume* 65, pp. 161–83.

Jeffrey, R. and R. C. Stalnaker 1994: 'Conditionals as Random Variables'. In Eells and Skyrms (1994), pp. 31–46.

Joyce, J. M. 1999: *The Foundations of Causal Decision Theory*. Cambridge Studies in Probability, Induction, and Decision Theory, Cambridge: Cambridge University Press.

Joyce, J. M. 2005: 'How Probabilities Reflect Evidence'. *Philosophical Perspectives* 19, pp. 153–78.

Kaufmann, S. 2004: 'Conditioning against the grain: Abduction and indicative conditionals'. *Journal of Philosophical Logic* 33, pp. 583–606.

Kaufmann, S. 2005: 'Conditional Predictions. A Probabilistic Account'. *Linguistics and Philosophy* 28, pp. 181–231.

Klinedinst, N. 2011: 'Quantified Conditionals and Conditional Excluded Middle'. *Journal of Semantics* 28, pp. 149–70.

Kölbel, M. 2000: 'Edgington on Compounds of Conditionals'. *Mind* 109, pp. 97–108.

Kratzer, A. 1979: 'Conditional Necessity and Possibility'. In Bäuerle, R., U. Egli, and A. von Stechow (eds.): *Semantics from different points of view*, Berlin, Heidelberg and New York: Springer, pp. 117–47.

Kratzer, A. 1981: 'The Notional Category of Modality'. In Eikmeyer, H. J. and H. Rieser (eds.): *Words, Worlds, and Contexts*, Berlin and New York: de Gruyter, pp. 38–74, an updated version is contained in Kratzer 2012: ch. 2.

Kratzer, A. 1986: 'Conditionals'. In Farley, A. M., P. Farley, and K. E. McCollough (eds.): *Papers from the Parasession on Pragmatics and Grammatical Theory*, Chicago: Chicago Linguistics Society, pp. 115–35, reprinted in von Stechow and Wunderlich 1991: ch. 30, 651–6. An updated version is contained in Kratzer 2012: ch. 4.

Kratzer, A. 2012: *Modals and Conditionals*. Oxford Studies in Theoretical Linguistics, Oxford: Oxford University Press.

Kripke, S. A. 1980: *Naming and Necessity*. Cambridge, MA: Harvard University Press.

Leitgeb, H. 2007: 'Beliefs in Conditionals vs. Conditional Beliefs'. *Topoi* 26, pp. 115–32.

Leitgeb, H. 2012a: 'A Probabilistic Semantics for Counterfactuals. Part A'. *The Review of Symbolic Logic* 5, pp. 26–84.

Leitgeb, H. 2012b: 'A Probabilistic Semantics for Counterfactuals. Part B'. *The Review of Symbolic Logic* 5, pp. 85–121.

Leslie, S.-J. 2009: ' "If", "Unless", and Quantification'. *Studies in Linguistics and Philosophy* 85, pp. 3–30.

Lewis, D. 1973: *Counterfactuals*. Oxford: Blackwell.

Lewis, D. 1975: 'Adverbs of Quantification'. In Keenan, E. L. (ed.): *Formal Semantics of Natural Language*, Cambridge: Cambridge University Press, pp. 3–15.

Lewis, D. 1976: 'Probabilities of Conditionals and Conditional Probabilities'. *Philosophical Review* 85, pp. 297–315. Page references to the reprint in Lewis 1986b.

Lewis, D. 1979: 'Counterfactuals and Time's Arrow'. *Noûs* 13, pp. 455–76, reprinted with a postscript in Lewis 1986b: ch. 2.

Lewis, D. 1980: 'A Subjectivist's Guide to Objective Chance'. In Jeffrey, R. C. (ed.): *Studies in Inductive Logic*, vol. 2, Berkeley and Los Angeles: University of California Press, reprinted with a postscript in Lewis 1986b: 83–133. Reprinted also in Harper et al. 1981: 267–97. Page references to the reprint in Lewis 1986b.

Lewis, D. 1981: 'Causal Decision Theory'. *Australasian Journal of Philosophy* 59, pp. 5–30.

Lewis, D. 1986a: *On the Plurality of Worlds*. Oxford: Blackwell.

Lewis, D. 1986b: *Philosophical Papers*, vol. 2. New York and Oxford: Oxford University Press.

Lewis, D. 1986c: 'Probabilities of Conditionals and Conditional Probabilities II'. *Philosophical Review* 95, pp. 581–9, reprinted in Jackson 1991: ch. 5.

Lewis, D. 1994: 'Humean Supervenience Debugged'. *Mind* 412, pp. 473–90.

Lycan, W. G. 2001: *Real Conditionals*. Oxford: Oxford University Press.

MacFarlane, J. 2003: 'Future Contingents and Relative Truth'. *The Philosophical Quarterly* 53, pp. 321–36.

MacFarlane, J. and N. Kolodny 2010: 'Ifs and Oughts'. *The Journal of Philosophy* 107, pp. 115–43.

McDermott, M. 1999: 'Counterfactuals and Access Points'. *Mind* 108, pp. 291–334.

McGee, V. 1985: 'A Counterexample to Modus Ponens'. *The Journal of Philosophy* 82, pp. 462–71.

McGee, V. 1989: 'Conditional Probabilities and Compounds of Conditionals'. *The Philosophical Review* 98, pp. 485–541.

McGee, V. 2000: 'To tell the truth about conditionals'. *Analysis* 60, pp. 107–11.

Moss, S. 2013: 'Subjunctive Credences and Semantic Humility'. *Philosophy and Phenomenological Research* 87, pp. 251–78.

Nolan, D. 1997: 'Impossible Worlds: A Modest Approach'. *Notre Dame Journal of Formal Logic* 38, pp. 535–72.

Phillips, I. 2007: 'Morgenbesser cases and closet determinism'. *Analysis* 67, pp. 42–9.

Pollock, J. L. 1976: *Subjunctive Reasoning*. Boston: Reidel.

Quine, W. V. O. 1950: *Methods of Logic*. Cambridge, MA: Harvard University Press.

Ramsey, F. P. 1926: 'Truth and Probability'. In Ramsey (1990), pp. 52–94.

Ramsey, F. P. 1929: 'General Propositions and Causality'. In Ramsey (1990), pp. 145–63.

Ramsey, F. P. 1990: *Philosophical Papers*. Cambridge: Cambridge University Press.

Rosenkranz, S. 2004: 'Fitch Back in Action Again?' *Analysis* 64, pp. 67–71.

Rothschild, D. 2013: 'Do Indicative Conditionals Express Propositions?' *Noûs* 47, pp. 49–68.

Schaffer, J. 2007: 'Deterministic Chance?' *The British Journal for the Philosophy of Science* 51, pp. 113–40.

Schröter, K. 1956: 'Theorie des bestimmten Artikels'. *Zeitschrift für mathematische Logik und Grundlagen der Mathematik* 2, pp. 37–56.

Schulz, M. 2010a: 'The Dynamics of Indexical Belief'. *Erkenntnis* 72, pp. 337–51.

Schulz, M. 2010b: 'Wondering What Might Be'. *Philosophical Studies* 149, pp. 367–86.

Schulz, M. 2011: 'Chance and Actuality'. *The Philosophical Quarterly* 61, pp. 105–29.

Schwarz, W. ms: 'Subjunctive Conditional Probability'. Ms, in progress.

Skyrms, B. 1980: *Causal Necessity*. New Haven and London: Yale University Press.

Skyrms, B. 1981: 'The Prior Propensity Account of Subjunctive Conditionals'. In Harper et al. (1981), pp. 259–65.

Slater, B. H. 1986: 'E-type Pronouns and ϵ-terms'. *Canadian Journal of Philosophy* 16, pp. 27–38.

Slote, M. 1978: 'Time in Counterfactuals'. *The Philosophical Review* 87, pp. 3–27.

Smith, Q. 2001: 'The Metaphysical Necessity of Natural Laws'. *Philosophica* 67, pp. 31–55.

Stalnaker, R. C. 1968: 'A Theory of Conditionals'. *Studies in Logical Theory* 2, pp. 98–112, reprinted in Jackson (1991: ch. 2).

Stalnaker, R. C. 1970: 'Probability and Conditionals'. *Philosophy of Science* 37, pp. 64–80, reprinted in Harper et al. (1981: 107–28).

Stalnaker, R. C. 1975: 'Indicative Conditionals'. *Philosophia* 5, pp. 269–86, reprinted in Jackson 1991: ch. 7.

Stalnaker, R. C. 1978: 'A Defense of Conditional Excluded Middle'. In Harper et al. (1981), pp. 87–104.

Stalnaker, R. C. 1984: *Inquiry*. Cambridge, MA and London: MIT Press.

Stalnaker, R. C. and R. Thomason 1968: 'Modality and Reference'. *Noûs* 2, pp. 359–72.

Starr, W. B. 2014: 'A Uniform Theory of Conditionals'. *Journal of Philosophical Logic* 43, pp. 1019–64.

Strawson, P. F. 1986: ' "If" and "⊃". ' In Grandy, R. E. and R. Warner (eds.): *Philosophical Grounds of Rationality*, Oxford: Clarendon Press, pp. 229–42, reprinted in Strawson 1997: ch. 9.

Strawson, P. F. 1997: *Entity and Identity and Other Essays*. Oxford: Clarendon Press.

Swanson, E. 2010: 'On Scope Relations between Quantifiers and Epistemic Modals'. *Journal of Semantics* 27, pp. 529–40.

Thau, M. 1994: 'Undermining and Admissibility'. *Mind* 103, pp. 491–503.

Thomason, R. 1970: 'Indeterminist time and truth-value gaps'. *Theoria* 36, pp. 264–81.

Tichy, P. 1978: 'A New Theory of Subjunctive Conditionals'. *Synthese* 37, pp. 433–57.

Tooley, M. 2002: 'Backward Causation and the Stalnaker-Lewis Approach to Counterfactuals'. *Analysis* 3, pp. 191–7.

van Fraassen, B. 1966: 'Singular Terms, Truth-Value Gaps and Free Logic'. *Journal of Philosophy* 63, pp. 481–95.

van Fraassen, B. 1976: 'Probabilities of Conditionals'. In Harper and Hooker (1976), pp. 261–301.

Veltman, F. 1996: 'Defaults in Update Semantics'. *Journal of Philosophical Logic* 25, pp. 221–61.

Veltman, F. 2005: 'Making Counterfactual Assumptions'. *Journal of Semantics* 22, pp. 159–80.

von Fintel, K. 1999: 'The Presupposition of Subjunctive Conditionals'. In Sauerland, U. (ed.): *The Interpretive Tract*, vol. 25, Cambridge, MA: MIT Working Papers in Linguistics, pp. 29–44.

von Fintel, K. 2001: 'Counterfactuals in a Dynamic Context'. In Kenstowicz, M. (ed.): *Ken Hale: A Life in Language*, Cambridge, MA: MIT Press, pp. 123–52.

von Fintel, K. 2003: 'Epistemic Modals and Conditionals Revisited'. Slides presented at the UMass Linguistics Colloquium. Available at http://web.mit.edu/fintel/www/umass-handout.pdf. Last accessed: 21 January 2016.

von Fintel, K. 2007: '*If*: The Biggest Little Word'. Slides presented at the Georgetown University Roundtable. Available at http://web.mit.edu/fintel/fintel-2007-if-gurt.pdf. Last accessed: 21 January 2016.

von Fintel, K. and S. Iatridou 2002: 'If and When If-Clauses Can Restrict Quantifiers'. Ms. Available at http://web.mit.edu/fintel/www/lpw.mich.pdf. Last accessed: 2 October 2016.

von Heusinger, K. 2004: 'Choice Functions and the Anaphoric Semantics of Definite NPs'. *Research on Language and Computation* 2, pp. 309–29.

von Stechow, A. and D. Wunderlich (eds.) 1991: *Semantics—An International Handbook of Contemporary Research*. Berlin and New York: de Gruyter.

von Wright, G. H. 1957: *Logical Studies*. London: Routledge and Kegan Paul.

Walters, L. 2009: 'Morgenbesser's Coin and Counterfactuals with True Components'. *Proceedings of the Aristotelian Society* 109, pp. 365–79.

Walters, L. 2011: 'Reply to Ahmed'. *Proceedings of the Aristotelian Society* 111, pp. 123–33.

Weatherson, B. 2001: 'Indicatives and Subjunctives'. *Philosophical Quarterly* 51, pp. 200–16.

Williams, J. R. G. 2008: 'Chances, Counterfactuals and Similarity'. *Philosophy and Phenomenological Research* 77.

Williams, J. R. G. 2009: 'Vagueness, Conditionals and Probability'. *Erkenntnis* 70, pp. 151–71.

Williams, J. R. G. 2010: 'Defending Conditional Excluded Middle'. *Noûs* 44, pp. 650–68.

Williams, J. R. G. 2012: 'Counterfactual Triviality: A Lewis-Impossibility Argument for Counterfactuals'. *Philosophy and Phenomenological Research* LXXXV, pp. 648–70.

Williamson, J. 2009a: 'Philosophies of Probability'. In Gabbay, D., P. Thagard, and J. Woods (eds.): *Handbook of the Philosophy of Mathematics*, North Holland, Oxford: Elsevier.

Williamson, J. 2011: 'Objective Bayesianism, Bayesian Conditionalization and Voluntarism'. *Synthese* 178, pp. 67–85.

Williamson, T. 1988: 'Bivalence and Subjunctive Conditionals'. *Synthese* 75, pp. 405–21.

Williamson, T. 1994: *Vagueness*. London: Routledge.

Williamson, T. 1996: 'Knowing and Asserting'. *Philosophical Review* 105, pp. 489–523.

Williamson, T. 1997: 'Knowledge as Evidence'. *Mind* 424, pp. 717–41.

Williamson, T. 1998: 'Conditionalizing on Knowledge'. *British Journal for the Philosophy of Science* 49, pp. 89–121.

Williamson, T. 2000: *Knowledge and Its Limits*. Oxford: Oxford University Press.

Williamson, T. 2006: 'Indicative versus subjunctive conditionals, congruential versus non-hyperintensional contexts'. In Sosa, E. and E. Villanueva (eds.): *Philosophy of Language, Philosophical Issues*, vol. 16, Oxford: Blackwell, pp. 310–33.

Williamson, T. 2007: 'How Probable is an Infinite Sequence of Heads?' *Analysis* 67, pp. 173–80.

Williamson, T. 2009b: 'Replies to Critics'. In Greenough and Pritchard (2009), pp. 279–384.

Williamson, T. forthcoming: 'Counterpossibles'. In Armour-Garb, B. and F. Kroon (eds.): *Philosophical Fictionalism*.

Woods, M. 1997: *Conditionals*. Oxford: Oxford University Press, with a commentary by Dorothy Edgington.

Zach, R. 2003: 'The practice of finitism. Epsilon calculus and consistency proofs in Hilbert's Program'. *Synthese* 137, pp. 211–59.

Index

The manufacturer's authorised representative in the EU for product safety is Oxford University Press España S.A. of El Parque Empresarial San Fernando de Henares, Avenida de Castilla, 2 - 28830 Madrid (www.oup.es/en or product.safety@oup.com). OUP España S.A. also acts as importer into Spain of products made by the manufacturer.
Printed and bound by CPI Group (UK) Ltd, Croydon, CR0 4YY
14/09/2026
02216288-0002